A Short Course in Cloud Physics

THIRD EDITION

INTERNATIONAL SERIES IN **NATURAL PHILOSOPHY**
VOLUME 113
General Editor: D. TER HAAR

A Short Course in Cloud Physics

THIRD EDITION

by

R. R. ROGERS

and

M. K. YAU

Department of Meteorology,
McGill University, Canada

An Imprint of Elsevier

Butterworth-Heinemann is an imprint of Elsevier

First published 1976
Second edition 1979
Reprinted 1979, 1983, 1985
Third Edition 1989
Reprinted 1996

 This book is printed on acid-free paper.

Library of Congress Cataloging-in-Publication Data
Rogers R. R. (Roddy Rhodes)
    A short course in cloud physics/R. R. Rogers, and M. K. Yau,
      p.   cm. (International series in natural philosophy; v.113)
    Bibliography: p. includes index
  1.   Cloud physics. I.   2. Yau, Man Kong II. Title III   Series
QC921.5.R63   1988 551.57'6-dc19   88-22709
621.384'12-dc20   CIP

ISBN 0-7506-3215-1

British Library Cataloguing-in-Publication Data
A catalogue record for this book is available from the British Library

The publisher offers special discounts on bulk orders of this book.
For information, please contact:
Manager of Special Sales
Elsevier Science
200 Wheeler Road
Burlington, MA 01803
Tel: 781-313-4700
Fax: 781-313-4802

For information on all Butterworth-Heinemann publications available, contact our
World Wide Web homepage at http://www.bh.com

10 9 8

Transferred to digital printing 2006

To the memory of our friend,

WALTER HITSCHFELD

Preface

This is an expanded and thoroughly revised edition of a book with the same title, published first in 1976, and in a second edition in 1979. As before, it is meant to serve as a text for a one-semester course in cloud physics. Although it grew out of lecture notes for an introductory graduate course in physical meteorology, it is also suitable for upper-level undergraduate students in the physical sciences or engineering.

Since the publication of the first edition, four other books have appeared, which are important additions to the material available for teaching physical meteorology at the graduate level. Foremost for cloud physics is the treatise by Pruppacher and Klett (1978), the most comprehensive book now available on cloud microphysics. The graduate student specializing in cloud physics will want to own a copy of this information-packed volume. J. T. Houghton's survey of atmospheric physics (second edition 1986) has only a short chapter on clouds, with emphasis on their radiative effects, but his book shows how to cover a broad range of topics with sweep and style, integrating atmospheric dynamics with atmospheric physics. The book on physical meteorology by H. G. Houghton (1985) includes mainly topics in radiation, cloud physics, and precipitation physics. Although less comprehensive than Pruppacher and Klett, its treatment of cloud physics is clear and authoritative. Finally, the book on cloud seeding by Dennis (1980) lucidly organizes this important part of cloud physics, giving it a coherence that had been lacking. We have learned from these books, and acknowledge their influence on this edition.

In general outline, this edition is very similar to the last. The first four chapters are on atmospheric thermodynamics, mainly to provide the necessary background for students new to meteorology. The most noticeable changes are a new chapter on observed properties of clouds (Chapter 5) and an entirely rewritten chapter on cloud modeling (Chapter 15). The section in Chapter 6 on aerosols has been expanded, and the subject of cloud-top mixing is discussed in several places. Symmetric instability is introduced in Chapter 3, allowing a broader interpretation of the mesoscale structure of precipitation. We made many small changes that are inconspicuous, but that improve the book. For example, the table of saturation vapor pressures was amended, in the light of recent critical data. We were almost successful in using SI units throughout. The main exceptions are for pressure, which still appears occasionally with units of millibars, and for the masses and fall speeds of cloud and precipitation particles, often given in CGS units. We were not able to standardize on radius or diameter as the appropriate measure of particle size. Also, in the interest of notational simplicity, the same symbol is occasionally used to represent different quantities. The student must therefore be vigilant, though the meaning of a symbol should be clear from the context. More problems are included than before, many of them new. A few are probably inappropriate for undergraduate students. Scarcely a page was unaltered in our efforts to improve clarity. We are grateful for the favor accorded the earlier editions, and hope this revision will also merit approval.

We are pleased to thank Dr. Robert F. Abbey, Jr., of the U.S. Office of Naval Research for his spirited encouragement. He provided the incentive to get the project started. Many other people generously assisted with details. Mr. S. G. Geotis of M.I.T. and Mr. Brooks Martner of the Wave Propagation Laboratory, NOAA, kindly provided us with examples of radar data. Satellite examples were supplied by Mr. John Lewis of the Canadian Atmospheric Environment Service and Mr. Tom Jaqua of NASA. We are grateful for their help. We thank Dr. George Isaac, also of the AES, for assistance in preparing the examples of microphysical measurements in Chapter 5. The radar examples from NCAR were provided by one of our graduate students, Ms. Miriam Blaskovic.

For permission to use figures appearing in books and journals, we thank the following publishers: Academic Press, American Geophysical Union, American Meteorological Society, Birkhäuser Verlag, Blackwell Scientific Publications, Cambridge University Press, Methuen & Co., MIT Press, Oxford University Press, Royal Meteorological Society, Swedish Geophysical Society, University of Chicago Press, Fried. Vieweg & Sohn.

We are indebted to our colleague, Dr. Henry G. Leighton, for his interest in the project and for his assistance with the new section on acidic precipitation. The influence of our associate from years past, Dr. David Robertson, is still apparent in the sections on ice nucleation and crystal growth. We were fortunate again to have the expert help of Miss Ursula Seidenfuss in preparing the illustrations.

We are deeply grateful to our new colleague, Dr. Gerhard W. Reuter, for his unflagging help during the final stages of the work. He wrote most of Chapter 5 and a substantial part of Chapter 15. The book in its present form owes much to his efforts. Other friends and colleagues too numerous to mention helped with advice and encouragement. To all of them, our thanks.

Montreal, April 1988   R. R. ROGERS
                                                         M. K. YAU

Contents

Introduction   xiii

1. Thermodynamics of Dry Air   1

    Atmospheric composition   1
    Equation of state for dry air   1
    The first law of thermodynamics   2
    Special processes   6
    Entropy   7
    Meteorological thermodynamic charts   8
    Problems   11

2. Water Vapor and its Thermodynamic Effects   12

    Equation of state for water vapor   12
    Clausius–Clapeyron equation   12
    Moist air : its vapor content   16
    Thermodynamics of unsaturated moist air   18
    Ways of reaching saturation   19
    Pseudoadiabatic process   21
    Adiabatic liquid water content   23
    Reversible saturated adiabatic process   25
    Problems   26

3. Parcel Buoyancy and Atmospheric Stability   28

    Hydrostatic equilibrium   28
    Dry adiabatic lapse rate   29
    Buoyant force on a parcel of air   30
    Stability criteria for dry air   30
    The pseudoadiabatic lapse rate   32
    Stability criteria for moist air   32
    Convective instability   33
    Horizontal restoring forces   35
    Geostrophic wind and geostrophic wind shear   36
    Slantwise displacement   38
    Symmetric instability   39
    Baroclinic instability   41
    Geopotential   41
    Problems   42

4. Mixing and Convection   **44**

  Mixing of air masses   44
  Convective condensation level   47
  Convection: elementary parcel theory   48
  Modification of the elementary theory   50
  Problems   57

5. Observed Properties of Clouds   **60**

  Sizes of clouds and cloud systems   60
  Microstructure of cumulus clouds   64
  Cloud droplet spectra   72
  Microstructure of stratus clouds   74
  Likelihood of ice and precipitation in clouds   75
  Microstructure of large continental storm clouds   77
  Problems   79

6. Formation of Cloud Droplets   **81**

  General aspects of cloud and precipitation formation   81
  Nucleation of liquid water in water vapor   84
  Atmospheric condensation nuclei   89
  Problems   96

7. Droplet Growth by Condensation   **99**

  Diffusional growth of a droplet   99
  The growth of droplet populations   105
  Some corrections to the diffusional growth theory   112
  Problems   119

8. Initiation of Rain in Nonfreezing Clouds   **121**

  Setting the stage for coalescence   122
  Droplet growth by collision and coalescence   124
  The Bowen model   131
  Statistical growth: the Telford model   134
  Statistical growth: the stochastic coalescence equation   137
  Condensation plus stochastic coalescence   143
  The effects of turbulence on collisions and coalescence   145
  Concluding remarks   147
  Problems   148

9. Formation and Growth of Ice Crystals   **150**

  Nucleation of the ice phase   150
  Experiments on heterogeneous ice nucleation   153
  Atmospheric ice nuclei   154

The ice phase in clouds 156
Diffusional growth of ice crystals 158
Further growth by accretion 163
The ice crystal process versus coalescence 166
Problems 169

10. Rain and Snow 170

Drop-size distribution 170
Drop breakup 172
Distribution of snowflakes with size 180
Aggregation and breakup of snowflakes 182
Precipitation rates 182
Problems 183

11. Weather Radar 184

Principles of radar 184
The radar equation 187
The weather radar equation 188
Relation of Z to precipitation rate 190
Radar displays and special techniques 191
Problems 193

12. Precipitation Processes 196

Stratiform precipitation 197
Showers 203
Precipitation theories 206
Mesoscale structure of rain 209
Precipitation efficiency 217
Acidic precipitation 218
Problems 220

13. Severe Storms and Hail 222

Life cycle of the thunderstorm cell 222
Severe thunderstorms 226
Precipitation production by thunderstorms 234
Hail growth 235
Problems 237

14. Weather Modification 240

Stimulation of rain and snow 241
Cloud dissipation 243
Hail suppression 243
Problems 244

15.   Numerical Cloud Models   **246**
   The governing equations   247
   One-dimensional models   249
   Two-dimensional models   253
   Three-dimensional models   259
   Model evaluation   266

References   269

Appendix   277

Answers to Selected Problems   278

Index   285

Other Titles in the Series in Natural Philosophy   291

Introduction

Cloud physics, as a branch of physical meteorology, may be defined as the science of clouds in the atmosphere. This definition accommodates a diverse body of knowledge, ranging from the optical properties of clouds to the chemistry of rainwater, from the physics of hail growth to the influence of clouds on the general circulation of the atmosphere. Within this wide range of subject matter, the core of cloud physics is usually regarded as the study of the formation of clouds and the development of precipitation. These subjects are central to radar meteorology, weather modification, severe storms research, and other major branches of physical meteorology. They must be understood, at least to some extent, as a background to almost any investigation of the physical effects of clouds.

For a cloud to form it is necessary that a large volume of air be chilled below its dew point. In the atmosphere the chilling is most often created by the approximately adiabatic expansion of ascending air, but it can be caused by radiative cooling or by the mixing of air masses with different temperatures and humidities. For precipitation to develop the chilling must continue, because this allows still more water vapor to condense. The rate of cooling, and the volume of air affected, determine the amount of condensation and hence the amount of precipitation that a cloud can produce. Though the rate of condensation is dictated by air motions, the formation of cloud droplets, their growth, and their interactions to produce precipitation particles are controlled by processes on a much smaller scale—a scale comparable in size to the individual cloud and precipitation particles. The core of cloud physics thus includes the study of phenomena on two widely different scales. With sizes ranging from tens of meters to hundreds of kilometers, we have air motions and the attendant thermodynamic changes; on the scale from centimeters down to micrometers we have the processes of droplet growth and interaction. As a specialization within the core, "cloud microphysics" refers to the study of the small scale processes. The large scale processes are usually called "cloud dynamics" or "cloud kinematics".

This book, which is intended for use in a first course in cloud physics, focuses on the core of the subject. Chapters 1–4 provide a sufficient background in meteorological thermodynamics to proceed with the study of cloud physics. Chapter 5, on the observed properties of clouds, is an overview intended to give the student some feel for the subject before embarking on the journey through theory. The central chapters, 6–10, are on the microphysics of clouds and precipitation. Here we have attempted, by focusing on fundamentals, to equip the student with what is needed to understand current journal articles or advanced books on these subjects. The later chapters move from microphysics to considerations of a larger scale. Chapter 11 is a brief introduction to meteorological radar, the instrument that has become essential in cloud physics research. Chapters 12 and 13, on precipitation processes and severe storms, show how radar observations are used to increase our understanding of cloud dynamics and microphysics. Chapter 14, another brief one, is on weather modification, a subject that has become somewhat detached from the core of cloud physics, but which has inspired so much activity in cloud physics that it should be included in even a short course. The most important recent advances in cloud dynamics have come from the computer simulation of clouds, the topic of the last chapter. Only through numerical modeling has it proven possible to understand the complex coupling between dynamics and microphysics, to make the detailed connections between theory and observations.

The book thus starts with the classical subject of thermodynamics and ends with numerical cloud modeling, an area of active current research in which new developments are rapidly emerging. We have tried throughout to be selective while covering the fundamentals, but ample references are included to enable the student to follow up details.

1

Thermodynamics of Dry Air

Atmospheric composition

Air is a mixture of several so-called permanent gases, a group of gases with variable concentrations, and different solid and liquid particles of variable concentrations. Nitrogen and oxygen account respectively for about 78% and 21% by volume of the atmosphere's permanent gases, with the remaining 1% consisting mainly of argon, but with trace amounts of neon, helium, and other gases. The composition of air is remarkably uniform, with the relative proportions of these permanent gases being essentially the same the world over and up to an altitude of 90 km.

The most abundant of the gases present in variable amounts are water vapor, carbon dioxide, and ozone. These gases strongly affect radiative transfer in the atmosphere. Water vapor is also of central importance in atmospheric thermodynamics.

The particles of solid and liquid material suspended in the air are called aerosols. Common examples are smoke, dust, and pollen. The water droplets and ice crystals of which clouds are composed are aerosols too, but they are usually classified with rain and other precipitation forms as *hydrometeors*, condensed forms of water in the atmosphere. Thermodynamics is concerned with the gases, but a select group of the aerosols called hygroscopic nuclei are crucial for the condensation of water in the atmosphere.

The approach in meteorology is to treat air as a mixture of two ideal gases: "dry air" and water vapor. This mixture is called moist air. The thermodynamic properties of moist air are determined by combining the separate thermodynamic behaviors of dry air and water vapor.

Equation of state for dry air

The equation of state for a perfect gas, or ideal gas law, expresses the relationship among pressure p, volume V, and temperature T of a gas in thermal equilibrium:

1

$$pV = CT, \tag{1.1}$$

with C a constant depending upon the particular gas.

The equation is reduced to standard form by employing Avogadro's Law, which states that at the same pressure and temperature, one mole of any gas occupies the same volume. Denoting this volume by v, we have

$$pv = C'T, \tag{1.2}$$

where C' is the same constant for all gases. It is called the *universal* gas constant, denoted by R^*, and equal to 8.314 J mole^{-1} K^{-1}.

Since an arbitrary volume $V = nv$, with n the number of moles, it follows from (1.2) that

$$pV = nR^*T. \tag{1.3}$$

Dividing by the mass M of the gas gives

$$\frac{pV}{M} = \frac{n}{M} R^*T.$$

But $V/M = \alpha$, the specific volume, and $n/M = 1/m$ where m denotes the molecular weight of the gas. Consequently (1.3) reduces to

$$p\alpha = R'T, \tag{1.4}$$

where $R' = R^*/m$ is called the *individual* gas constant.

It is possible to calculate the effective molecular weight of dry air by suitably averaging the molecular weights of the nitrogen, oxygen, and trace gases of which it is composed. This turns out to be 28.96 g/mole. Accordingly, the individual gas constant for dry air is

$$R' = 287 \text{ J kg}^{-1} \text{ K}^{-1}.$$

Over the meteorological range of temperature and pressure, (1.4) describes the behavior of dry air with sufficient accuracy for most purposes.

For consistent SI units in (1.4), pressure is in Pa, specific volume in m^3/kg, and temperature in K. It is conventional in meteorology, however, to measure pressure in kilopascals (1 kPa $= 10^3$ Pa) or millibars (1 mb $= 10^3$ dynes/cm$^2 = 10^2$ Pa $= 1$ hPa $= 10^{-1}$ kPa). Temperature is often measured in degrees Celsius, related to degrees Kelvin by T (in K) $= T$ (in °C) $+ 273.15$.

The first law of thermodynamics

The first law is a statement of two empirical facts:

1. Heat is a form of energy.
2. Energy is conserved.

The first of these is called Joule's Law, and expresses the mechanical equivalent of heat as

$$1 \text{ cal} = 4.1868 \text{ J}. \tag{1.5}$$

The second of these empirical facts may be stated in algebraic form:

$$dQ = dU + dW. \tag{1.6}$$

Of the total amount of heat added to a gas, dQ, some may tend to increase the internal energy of the gas by amount dU, and the remainder will cause work to be done by the gas in the amount dW. It is generally more useful to express this relation for a unit mass of gas, for which (1.6) becomes

$$dq = du + dw. \tag{1.7}$$

We examine first the work term in (1.7). Consider a parcel of gas with volume V and surface area A, as illustrated in Fig. 1.1. The change in volume associated with a small incremental linear expansion dn is

$$dV = A\,dn.$$

But $p = F/A$, where F is the force exerted by the gas, so that

$$p\,dV = F\,dn. \tag{1.8}$$

The work done by the gas in expanding is $dW = F\,dn$. Consequently (1.8) may be written

$$dW = p\,dV.$$

The work done per unit mass of gas (specific work) is

$$dw = p\,d\alpha. \tag{1.9}$$

In general, the specific work done in a finite expansion from α_1 to α_2 is

$$\int dw = \int_{\alpha_1}^{\alpha_2} p\,d\alpha.$$

This integration may be visualized with the help of a thermodynamic diagram.

A thermodynamic diagram is a chart whose coordinates are variables of state. A given equilibrium thermodynamic state of a gas may be

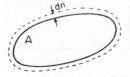

FIG. 1.1. Expanding parcel of gas.

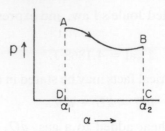

FIG. 1.2. Thermodynamic chart representing work done in expansion.

represented by a point on such a chart. As a gas goes through successive equilibrium states (for example in response to heating or to an external force), it traces out a path on a thermodynamic diagram.

The work done by a gas in expanding is readily illustrated on a chart with coordinates of pressure versus specific volume, as in Fig. 1.2. In the example shown the gas expands from initial state $A(p_1, \alpha_1)$ to final state $B(p_2, \alpha_2)$. The specific work done is represented by the area $ABCD$. There are actually any number of possible equilibrium paths from A to B, depending upon whether heat is added to or taken from the gas, and at what point during the process this heat transfer occurs. The work depends on the path of integration, which is another way of saying $dw = pd\alpha$ is not an exact differential.

Of special interest in thermodynamic theory are *cyclic* processes, in which the gas undergoes a continuous series of changes in state, but ends up with the same thermodynamic coordinates it had initially. One such cyclic process is pictured in Fig. 1.3 on a p,α-diagram. The gas starts at state A and proceeds to state B along the indicated curve. As before, the area under this curve gives the work done by the gas in expanding from α_1 to α_2. Next the gas is compressed and made to return to state A along the lower curve. In this step of the process work is done *on* the gas. The net work done by the gas in the complete cyclic process is given by the hatched area. Note that if the process had taken place in the opposite sense, with arrows reversed, the hatched area would stand for net work

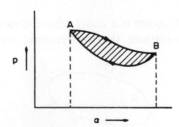

FIG. 1.3. Cyclic process.

done *on* the gas. The net work in a cyclic process is described mathematically by an integral over the closed path,

$$\oint dw = \oint p \, d\alpha.$$

For integrands that are exact differentials, any such cyclic integration yields zero, because the integral over an exact differential depends only on the limits of integration. This is not the case in general for dw, as already explained.

Next we consider the du term in (1.7). For an ideal gas, any increase in internal energy appears as an increase in temperature. The temperature change is proportional to the amount of heat added according to

$$dT = \frac{1}{c} \, dq, \qquad (1.10)$$

where c is called the specific heat capacity and is measured, for example, in $J \, kg^{-1} \, K^{-1}$ or $cal \, g^{-1} \, K^{-1}$. For a gas, c is not constant but depends upon whether work is done while the heat is added. If no work is done, $d\alpha = 0$ from (1.9) and for the specific heat we write

$$c_v = \left(\frac{dq}{dT} \right)_{\alpha}, \qquad (1.11)$$

known as the specific heat at constant volume.

Another case of interest is the addition of heat with pressure held constant, for which process the specific heat is given as

$$c_p = \left(\frac{dq}{dT} \right)_{p}, \qquad (1.12)$$

and called the specific heat at constant pressure.

Evidently $c_p > c_v$, because in a constant pressure process some of the added heat will be used in the work term $p \, d\alpha$, while in the constant volume process all added heat goes toward increasing T. For dry air,

$$c_p = 1005 \, J \, kg^{-1} \, K^{-1} = 0.240 \, cal \, g^{-1} \, K^{-1}$$
$$c_v = 718 \, J \, kg^{-1} \, K^{-1} = 0.171 \, cal \, g^{-1} \, K^{-1}.$$

Of the total head added, the amount that goes into the internal energy is

$$du = c_v dT \qquad (1.13)$$

and the remainder goes into the work term. Thus the general expression for the conservation of energy is

$$dq = c_v dT + p \, d\alpha. \qquad (1.14)$$

Special processes

By differentiating (1.4), we obtain

$$pd\alpha + \alpha dp = R'dT \qquad (1.15)$$

as a differential equation relating changes of pressure, specific volume, and temperature under conditions of thermodynamic equilibrium. Combining (1.15) and (1.14),

$$dq = (c_v + R')dT - \alpha dp.$$

But

$$c_p = \left(\frac{dq}{dT}\right)_p = c_v + R',$$

so that

$$dq = c_p dT - \alpha dp, \qquad (1.16)$$

which may be used instead of (1.14) as an expression of the first law.

Certain special processes are now defined using these equations.

(a) Isobaric process: $dp = 0$

$$dq = c_p dT = \left(\frac{c_p}{c_v}\right)c_v dT = \left(\frac{c_p}{c_v}\right)du. \qquad (1.17)$$

(b) Isothermal process: $dT = 0$

$$dq = -\alpha dp = pd\alpha = dw. \qquad (1.18)$$

(c) Isochoric process: $d\alpha = 0$

$$dq = c_v dT = du. \qquad (1.19)$$

(d) Adiabatic process: $dq = 0$

$$c_p dT = \alpha dp \qquad (1.20)$$

or

$$c_v dT = -pd\alpha. \qquad (1.21)$$

The adiabatic process is of special significance because many of the temperature changes that take place in the atmosphere can be approximated as adiabatic. From (1.20) and the equation of state,

$$c_p dT = R'T\frac{dp}{p}, \qquad (1.22)$$

which may be integrated to give

$$\left(\frac{T}{T_0}\right) = \left(\frac{p}{p_0}\right)^k, \tag{1.23}$$

where $k = R'/c_p = (c_p - c_v)/c_p = 0.286$.

The result (1.23) is called Poisson's equation for adiabatic processes. It is possible to derive expressions equivalent to (1.23) relating any two of the thermodynamic variables pressure, temperature, and specific volume.

A fourth thermodynamic variable, called the potential temperature, is defined on the basis of (1.23). It is denoted by θ and defined by

$$\left(\frac{T}{\theta}\right) = \left(\frac{p}{100 \text{ kPa}}\right)^k$$

or

$$\theta = T\left(\frac{100 \text{ kPa}}{p}\right)^k \tag{1.24}$$

and may be interpreted as the temperature that a parcel of air would have if, starting with temperature T at pressure p, it were subjected to an adiabatic compression or expansion to a final pressure of 100 kPa. The potential temperature is called a variable of state, because it is expressible in terms of the state variables p and T. In any adiabatic process, θ is a constant. We say that potential temperature is a *conservative* property in adiabatic processes.

Entropy

The second law of thermodynamics implies the existence of another variable of state, called entropy, which may be defined by the equation

$$d\phi = \frac{dq}{T}, \tag{1.25}$$

where $d\phi$ is the increase in (specific) entropy accompanying the addition of heat dq to a unit mass of gas at temperature T. It follows from (1.16) that

$$d\phi = \frac{1}{T}[c_p dT - \alpha dp] = c_p \frac{dT}{T} - R' \frac{dp}{p} = c_p\left[\frac{dT}{T} - k\frac{dp}{p}\right]$$

$$= c_p \frac{d\theta}{\theta}. \tag{1.26}$$

Integration gives

$$\phi = c_p \ln \theta + \text{const.}, \tag{1.27}$$

which associates entropy with potential temperature. It is evident from the defining equation (1.25) that adiabatic processes ($dq = 0$) are also isentropic processes.

Meteorological thermodynamic charts

(a) Stüve diagram

The Stüve diagram (or simply "adiabatic chart") is a thermodynamic diagram based on the adiabatic equation (1.24). This equation shows that, for a given value of θ, there is a linear relation between T and p^k. Consequently, adiabatic processes follow straight line paths on a thermodynamic diagram with coordinates of T versus p^k.

This kind of chart is convenient for depicting adiabatic processes in the atmosphere. A line along which θ = const. is called an *adiabat*. Figure 1.4 is a schematic diagram of a Stüve chart, illustrating the working coordinates of pressure and temperature, and also the appearance of isobars, adiabats, and isotherms.

(b) Emagram

So-called *true* thermodynamic diagrams are ones on which area is proportional to energy. Thus a p,α-diagram is a true thermodynamic diagram, because the area in any closed contour is proportional to the work done in a cyclic process defined by the contour.

In meteorology the state variables most frequently employed to describe the air are pressure and temperature. It is possible to construct a true thermodynamic diagram with coordinates of p and T on the basis of (1.9) and (1.15). We have

$$dw = pd\alpha = R'dT - \alpha dp$$

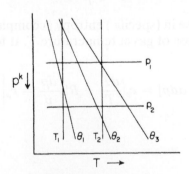

FIG. 1.4. Stüve diagram.

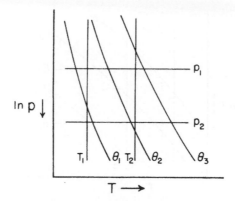

FIG. 1.5. Schematic emagram.

and, for a cyclic process,

$$\oint dw = \oint R'dT - \oint R'T\frac{dp}{p}. \tag{1.28}$$

But $R'dT$ is an exact differential which integrates to zero, so that the work done reduces to

$$\oint dw = -R' \oint Td(\ln p). \tag{1.29}$$

This result indicates that a chart with coordinates of T versus $\ln p$ has the property of a true thermodynamic diagram. Such a chart is called an emagram, an abbreviation for energy-per-unit-mass diagram, and is illustrated schematically in Fig. 1.5.

(c) Tephigram

From the defining equation of entropy, it follows that the total heat added in a cyclic process is

$$\oint dq = \oint Td\phi = c_p \oint Td(\ln \theta). \tag{1.30}$$

Consequently a chart with coordinates of T versus ϕ, or equivalently T versus $\ln \theta$, has the required area–energy relation of a true thermodynamic diagram. Such a chart is called a tephigram, standing for T,ϕ-diagram, and is shown schematically in Fig. 1.6.

The tephigram is usually rotated so that the isobars end up more or less horizontal with pressure decreasing upwards on the chart. Figure 1.7 illustrates a chart with this orientation. It is based on the Canadian

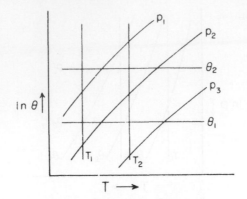

FIG. 1.6. Schematic tephigram.

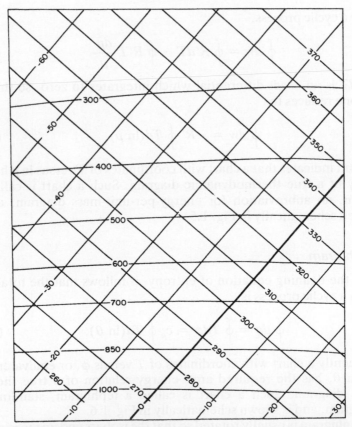

FIG. 1.7. Skeleton of a tephigram. Isobars are approximately horizontal, labeled in mb. Isotherms in deg C go upwards and to the right. Dry adiabats are normal to the isotherms and are labeled according to potential temperature (deg K).

meteorological service tephigram and will be used in the examples that follow.

Problems

1.1. The following table gives the approximate percentages by mass of the atmosphere's main permanent gases. Using these data, show that the effective molecular weight of dry air is 28.96 g/mole.

Gas	Mol. wt.	Mass, %
Nitrogen	28.016	75.57
Oxygen	32.000	23.15
Argon	39.944	1.28

1.2. A unit mass of dry air undergoes a Carnot cycle consisting of the following steps:

(a) adiabatic compression from 60 kPa and 0°C to a temperature of 25°C;
(b) isothermal expansion to a pressure of 70 kPa;
(c) adiabatic expansion to a temperature of 0°C;
(d) isothermal compression to the original pressure of 60 kPa.

Calculate the work done by the air in this process. Confirm your result using the tephigram or another meteorological thermodynamic chart.

1.3. A polytropic process for an ideal gas is one in which pressure and volume are related by $pV^n = $ const., where n is a constant. It is a generalization of the special processes considered earlier. Thus $n = 0$ defines an isobaric process, $n = c_p/c_v$ an adiabatic process, $n = 1$ an isothermal process, and $n - \infty$ an isochoric process. Suppose 1 kg of dry air at 280 K and 100 kPa undergoes a polytropic expansion in which the pressure falls to 70 kPa and the potential temperature increases by 10 K. Solve for

(a) the value of n;
(b) the change in internal energy of the air;
(c) the work done by the air;
(d) the heat absorbed by the air.

1.4. The coordinates x and y of any thermodynamic chart are of the form $x = x(p, \alpha)$ and $y = y(p, \alpha)$. For example, an emagram has $x = p\alpha/R' = T$ and $y = -R' \ln p$. A true thermodynamic diagram is one whose coordinates (x, y) may be transformed to (p, α) with a Jacobian of unity. That is, a given thermodynamic chart is a true thermodynamic diagram if

$$J\left(\frac{x, y}{p, \alpha}\right) = 1.$$

(a) Prove that a Stüve diagram is not a true thermodynamic diagram.
(b) How would you construct a true thermodynamic diagram with working coordinates of temperature and density? Indicate such a diagram schematically and sketch the approximate appearance of isobars and dry adiabats.

1.5. The specific enthalpy of a gas is defined by $h = u + p\alpha$.

(a) Prove that dh for an ideal gas is an exact differential.
(b) Calculate the change in enthalpy of a unit mass of dry air as it is compressed adiabatically from an initial pressure of 70 kPa and temperature of 10°C to a final pressure of 100 kPa.

1.6. A 200-gram sample of dry air is heated isobarically. Its entropy increases by 19.2 J/K and the work done by expansion is 1.61×10^3 J. Solve for the final temperature of the air.

2

Water Vapor and its Thermodynamic Effects

Equation of state for water vapor

Unlike other atmospheric constituents, water appears in all three phases, solid, liquid, and vapor. In the vapor phase water in the atmosphere behaves as an ideal gas to a good approximation. Its equation of state is

$$e = \varrho_v R_v T, \qquad (2.1)$$

where e = vapor pressure, ϱ_v = vapor density, and R_v = individual gas constant for water vapor ($461.5 \ \mathrm{J \ kg^{-1} \ K^{-1}}$). This equation sometimes appears in the form

$$e = \varrho_v \frac{R'}{\varepsilon} T, \qquad (2.2)$$

where $\varepsilon = R'/R_v = m_v/m = 0.622$.

Clausius–Clapeyron equation

Consider a closed and thermally insulated container partly filled with water as shown in Fig. 2.1. Molecules from the surface layer of water are in agitation and some break away as vapor molecules. On the other hand some of the vapor molecules collide with the surface and stick. Condensation and evaporation thus take place simultaneously. For a given

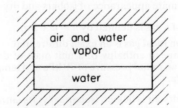

air and water vapor

water

FIG. 2.1. Vapor in equilibrium with liquid surface.

temperature an equilibrium condition will eventually be reached when the two processes have the same rate. Then the temperature of the air and vapor equals that of the liquid and there is no net transfer of molecules from one phase to the other. The space above the liquid is then said to be *saturated* with water vapor. The partial pressure due to the water vapor in this condition is called the saturation vapor pressure. It is found that the saturation vapor pressure depends only on temperature, and this functional dependence is described by an important differential equation that we will now derive.

Heat is required to change phase from liquid to vapor because the kinetic energy of the vapor molecules exceeds that of liquid molecules at the same temperature. We denote by L the heat required to convert a unit mass of liquid to vapor, with pressure and temperature held constant. This is the latent heat of vaporization. For this transition, from phase 1 (liquid) to phase 2 (vapor),

$$L = \int_{q_1}^{q_2} dq = \int_{u_1}^{u_2} du + \int_{a_1}^{a_2} p\,da = u_2 - u_1 + e_s(a_2 - a_1), \quad (2.3)$$

where e_s denotes the saturation vapor pressure, which is constant throughout the process. Because the temperature is also constant, we may write

$$L = T \int_{q_1}^{q_2} \frac{dq}{T} = T(\phi_2 - \phi_1). \quad (2.4)$$

Equating results, we find that

$$u_1 + e_s a_1 - T\phi_1 = u_2 + e_s a_2 - T\phi_2, \quad (2.5)$$

which shows that this particular combination of thermodynamic variables remains constant in an isothermal, isobaric change of phase. This combination is called the Gibbs function of the system and is denoted by G. Thus, for phase 1,

$$G_1 = u_1 + e_s a_1 - T\phi_1 \quad (2.6)$$

and (2.5) may be written as $G_1 = G_2$.

Though it is constant in the phase transition, the Gibbs function varies with temperature and pressure, and its dependence on these variables may be determined by differentiation:

$$dG = du + e_s da + a\,de_s - T\,d\phi - \phi\,dT. \quad (2.7)$$

But $du + e_s da = dq = T\,d\phi$, and (2.7) reduces to

$$dG = a\,de_s - \phi\,dT. \quad (2.8)$$

Because G is the same for both phases, $dG_1 = dG_2$ and (2.8) implies

$$\frac{de_s}{dT} = \frac{\phi_2 - \phi_1}{\alpha_2 - \alpha_1} = \frac{L}{T(\alpha_2 - \alpha_1)}. \tag{2.9}$$

This result expresses the change of saturation vapor pressure with temperature and is known as the Clausius–Clapeyron equation. Under ordinary atmospheric conditions $\alpha_2 \gg \alpha_1$ and water vapor behaves as an ideal gas. Then (2.9) reduces to

$$\frac{de_s}{dT} = \frac{L}{T\alpha_2} = \frac{Le_s}{R_v T^2}. \tag{2.10}$$

At temperatures colder than 0°C, (2.10) describes the saturation vapor pressure of supercooled liquid water. Ice may also exist and be in equilibrium with the vapor at subfreezing temperatures. The change with temperature of the saturation vapor pressure of ice is given by the Clausius–Clapeyron equation, except with L in (2.10) replaced by L_s, the latent heat of sublimation. At temperatures warmer than 0°C, only liquid water can be in equilibrium with the vapor.

As a first approximation, the Clausius–Clapeyron equation can be integrated by regarding the latent heat as constant. The result is

$$\ln \frac{e_s(T)}{e_{s0}} = \frac{L}{R_v} \left(\frac{1}{T_0} - \frac{1}{T} \right), \tag{2.11}$$

where e_{s0} is the value of saturation vapor pressure at temperature T_0, a constant of integration that must be determined by experiment. It is found that $e_{s0} = 611$ Pa at $T_0 = 0°C$. The latent heat of vaporization near 0°C is approximately 2.50×10^6 J/kg. Substituting these values in (2.11) gives, as an approximation for the saturation vapor pressure over water,

$$e_s(T) = Ae^{-B/T}, \tag{2.12}$$

where $A = 2.53 \times 10^8$ kPa and $B = 5.42 \times 10^3$ K.

The latent heat of vaporization depends weakly on temperature, changing by about 6% over the temperature range from $-30°C$ to $+30°C$. This dependence may be inferred from (2.3). We start by noting that in this equation $\alpha_2 \gg \alpha_1$, and $e_s\alpha_2 = R_v T$. Then, differentiating with respect to T, we find

$$\frac{dL}{dT} = c_{vv} - c + R_v = c_{pv} - c \tag{2.13}$$

where $c_{vv} = du_2/dT$ is the specific heat capacity of water vapor at constant volume; $c = du_1/dT$ is the specific heat capacity of liquid water; and $c_{pv} = R_v + c_{vv}$ is the specific heat of water vapor at constant pressure. Regarding the specific heats as constants, we may integrate (2.13) and write

$$L(T) = L_0 - (c - c_{pv})(T - T_0), \tag{2.14}$$

where $L_0 = L(T_0)$ is a constant of integration. A more accurate approximation than (2.12) for $e_s(T)$ can then be obtained by substituting (2.14) in the Clausius–Clapeyron equation and integrating (see problem 2.5).

The specific heat capacities also depend on temperature and pressure, although very weakly. At the saturation vapor pressure, for example, the specific heat at constant pressure c_{pv} increases with increasing temperature and is about 2% larger at $+30°C$ than at $-30°C$. For many purposes this variation can be neglected, and the value of c_{pv} can be taken as 1870 J kg^{-1} K^{-1}. A good approximation for the specific heat at constant volume is 1410 J kg^{-1} K^{-1}. For liquid water, the specific heat is within 1% of 1 cal g^{-1} K^{-1} = 4187 J kg^{-1} K^{-1} at temperatures warmer than 0°C, but as the temperature falls below 0°C this quantity gradually increases to a value approximately 8% larger at $-30°C$.

A first approximation for the saturation vapor pressure e_i of ice is (2.12), with $A = 3.41 \times 10^9$ kPa and $B = 6.13 \times 10^3$ K. This approximation follows from replacing L in (2.11) by L_s, the latent heat of sublimation, and using 2.83×10^6 J kg^{-1} as the value of L_s and 611 Pa as the value of e_i at 0°C. (Experiments show that e_s and e_i are equal to within three significant figures at 0°C.)

Comparing (2.11) for water and ice shows that, at a given subfreezing temperature,

$$\frac{e_s(T)}{e_i(T)} = \exp\left\{ \frac{L_f}{R_v T_0} \left(\frac{T_0}{T} - 1 \right) \right\} \tag{2.15}$$

where $L_f = L_s - L$ is the latent heat of fusion of water. Numerically, in the vicinity of 0°C, a good approximation to this equation is

$$\frac{e_s(T)}{e_i(T)} \approx \left(\frac{273}{T} \right)^{2.66}, \tag{2.16}$$

where T is in K. These relations indicate that the saturation vapor pressure of water exceeds that of ice for all temperatures below 273 K and that the ratio e_s/e_i steadily increases as the temperature decreases. Any atmosphere saturated with respect to water is supersaturated relative to ice; the degree of supersaturation increases with the supercooling.

Although the relationships given thus far are adequate for many purposes, accurate and precise values of the saturation vapor pressure are needed for some applications. Table 2.1 lists the accepted standard values of e_s and e_i over the range from $-40°C$ to $+40°C$. Also included are values of L and L_s. The vapor pressures were obtained by integrating the Clausius–Clapeyron equation, taking into account the most complete information on the dependence on temperature of the latent heats and

TABLE 2.1. *Saturation Vapor Pressures Over Water and Ice,*
and Latent Heats of Condensation and Sublimation

T(°C)	e_s (Pa)	e_i (Pa)	L (J/g)	L_s (J/g)
-40	19.05	12.85	2603	2839
-35	31.54	22.36		2839
-30	51.06	38.02	2575	2839
-25	80.90	63.30		2838
-20	125.63	103.28	2549	2838
-15	191.44	165.32		2837
-10	286.57	259.92	2525	2837
-5	421.84	401.78		
0	611.21	611.15	2501	2834
5	872.47		2489	
10	1227.94		2477	
15	1705.32		2466	
20	2338.54		2453	
25	3168.74		2442	
30	4245.20		2430	
35	5626.45		2418	
40	7381.27		2406	

specific heats, and fitting the curves to the points where definitive experimental values are known. The values of e_s for $T \geqslant 0°C$ are taken from Wexler (1976); those for e_i are from Wexler (1977). There is some uncertainty about the values of e_s for $T < 0°C$ owing to a lack of experimental data. The entries here were obtained by extrapolating Wexler's formula to temperatures below 0°C. The values of latent heat were taken from the 6th edition (R. J. List, 1951) of the Smithsonian Meteorological Tables. Bolton (1980) has shown that the tabulated data on $e_s(T)$ for water can be fitted to within 0.1% over the temperature range $-30°C \leqslant T \leqslant 35°C$ by the empirical formula

$$e_s(T) = 6.112 \exp\left(\frac{17.67T}{T + 243.5}\right), \tag{2.17}$$

where e_s is in mb and T is in degrees C.

Moist air: its vapor content

Atmospheric air is a mixture of dry air and water vapor. There are different ways to describe the vapor content, depending on the application.

(a) Vapor pressure e, the partial pressure of the water vapor.
(b) Vapor density ϱ_v, also called absolute humidity, defined by (2.1).
(c) Mixing ratio w, defined as the mass of water vapor per unit mass of dry air.

$$w = M_v/M_d = \varrho_v/\varrho_d.$$

From the equation of state, $\varrho_v = e/R_v T$ and $\varrho_d = (p - e)/R'T$ so that

$$w = \varepsilon \frac{e}{p - e} \approx \varepsilon \frac{e}{p}. \tag{2.18}$$

(d) Specific humidity q, the mass of water vapor per unit mass of moist air.

$$q = \frac{\varrho_v}{\varrho} = \frac{\varrho_v}{\varrho_d + \varrho_v} = \varepsilon \frac{e}{p - (1 - \varepsilon)e} \approx \varepsilon \frac{e}{p}. \tag{2.19}$$

The saturation mixing ratio and saturation specific humidity, denoted by w_s and q_s, are defined by (2.18) and (2.19) by formally replacing e by e_s. Because $e_s = e_s(T)$, w_s and q_s are functions of temperature and pressure only, and do not depend on the vapor content of the air. All meteorological thermodynamic charts contain "vapor lines", which are usually isopleths of w_s.

(e) Relative humidity f, the ratio of the mixing ratio to its saturation value at the same temperature and pressure.

$$f = \frac{w}{w_s(p, T)} \approx \frac{e}{e_s}. \tag{2.20}$$

The relative humidity is usually expressed in per cent.

(f) Virtual temperature T_v, the temperature of dry air having the same density as a sample of moist air at the same pressure.

For a sample of air of volume V, having total pressure p and vapor pressure e,

$$p = p_d + e = \varrho_d \frac{R^*}{m_d} T + \varrho_v \frac{R^*}{m_v} T$$

$$= \frac{R^* T}{V} \left[\frac{M_d}{m_d} + \frac{M_v}{m_v} \right]$$

$$= \varrho R^* T \left[\frac{M_d}{m_d} + \frac{M_v}{m_v} \right] \frac{1}{M_d + M_v}$$

$$= \varrho R' T \left[\frac{1 + w/\varepsilon}{1 + w} \right].$$

This result indicates that the equation of state for dry air may be applied to moist air if we include the correction factor in brackets. The virtual temperature is introduced to include this correction factor.

$$T_v = T \left[\frac{1 + w/\varepsilon}{1 + w} \right] \approx T[1 + 0.6w]. \tag{2.21}$$

Thermodynamics of unsaturated moist air

(a) Gas constant

The equation of state for dry air can be applied to moist air if we replace T by T_v. Thus

$$p\alpha = R'T_v \tag{2.22}$$

is a general equation of state applicable to dry or moist air. Frequently the difference between actual and virtual temperature is small and may be neglected.

Alternatively the equation of state for moist air can be written

$$p\alpha = R_m T, \tag{2.23}$$

where R_m is the individual gas constant for moist air, which must depend on the mixing ratio according to

$$R_m = R'[1 + 0.6w]. \tag{2.24}$$

Obviously (2.22) and (2.23) are equivalent; it only depends on where you choose to apply the correction factor.

(b) Specific heat

To determine c_{vm}, the specific heat at constant volume for moist air, consider the addition of heat to a sample of air consisting of one kilogram of dry air and w kilograms of water vapor.

$$(1 + w)dq = c_v dT + wc_{vv}dT,$$

where c_v is the specific heat of dry air and c_{vv} is the specific heat of the vapor. This shows that

$$c_{vm} = \left(\frac{dq}{dT}\right) = c_v\left[\frac{1 + wr}{1 + w}\right],$$

where

$$r = c_{vv}/c_v = 1410/718 = 1.96.$$

Thus

$$c_{vm} \approx c_v[1 + w]. \tag{2.25}$$

The same procedure may be employed to show that the specific heat at constant pressure for moist air is

$$c_{pm} \approx c_p[1 + 0.9w]. \tag{2.26}$$

Combining (2.24) and (2.26), the exponent in the Poisson equation

(1.23) for adiabatic processes in moist air is found to be

$$\frac{R_m}{c_{pm}} \approx k[1 - 0.2w].\tag{2.27}$$

Because w is of the order 10^{-2} or less, the correction factors in (2.24)–(2.27) may often be neglected.

Ways of reaching saturation

A sample of moist air may undergo several processes that lead to saturation. Some of these processes are of theoretical importance, and introduce certain new temperatures that reflect the moisture content of the air.

(a) Dew point temperature T_d, defined as the temperature to which moist air must be cooled, with p and w held constant, for it to reach saturation with respect to water. (The frost point temperature is defined similarly, except for saturation relative to ice.) Clearly the saturation mixing ratio at the dew point equals the mixing ratio of the moist air: $w = w_s(p, T_d)$. An analytical approximation for the dew point follows from (2.12):

$$T_d = T_d(w, p) = B/\ln (A\varepsilon/wp).\tag{2.28}$$

Graphical determination of the dew point is illustrated in Fig. 2.2.

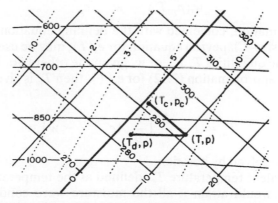

FIG. 2.2. Temperature, dew point, and isentropic condensation temperature, indicated on a tephigram. In the example shown the sample of air at 10°C, 900 mb, is assumed to have a mixing ratio of 5 g/kg. Its dew point, found from the intersection of the 900 mb isobar and the 5 g/kg vapor line, is 2.2°C. Its isentropic condensation point, found from the intersection of the adiabat through (T, p) with the 5 g/kg vapor line, is at 0.7°C and approximately 800 mb.

(b) Wet-bulb temperature T_w, defined as the temperature to which air may be cooled by evaporating water into it at constant pressure, until saturation is reached. (Note that w is not held constant, so that $T_d \neq T_w$ in general.)

Consider a sample of moist air consisting of one kilogram of dry air and w kilograms of water vapor. The first law of thermodynamics for this sample in an isobaric process is (from 2.26)

$$dq = c_p dT[1 + 0.9w].$$

Associated with the evaporation of mass dw of water is a heat loss given by

$$(1 + w)dq = -Ldw.$$

Consequently

$$c_p dT = -Ldw\left(\frac{1}{1 + w}\right)\left(\frac{1}{1 + 0.9w}\right) \approx -Ldw[1 - 1.9w].$$

It is often satisfactory to neglect the correction factor, so that

$$c_p dT = -Ldw, \tag{2.29}$$

which approximately describes the wet-bulb process. Neglecting the weak dependence of L on temperature, this equation may be integrated to yield

$$\frac{T - T_w}{w_s(p, T_w) - w} = \frac{L}{c_p}. \tag{2.30}$$

This approximation, combined with the defining equation for $w_s(p, T)$ and the Clausius–Clapeyron equation for $e_s(T)$, may be used to solve for any one of the quantities T, T_w, or w, given the other two. For example, if we use the approximation (2.12) for $e_s(T)$, then T_w is given in terms of T and w by

$$T_w = T - \frac{L}{c_p}\left(\frac{\varepsilon}{p} A e^{-B/T_w} - w\right), \tag{2.31}$$

which may be solved by iteration.

(c) Equivalent temperature T_e, defined as the temperature a sample of moist air would attain if all the moisture were condensed out at constant pressure. An expression for T_e follows from (2.30) if we set $w_s = 0$ (the final mixing ratio) and $T_w = T_e$. Thus

$$T_e = T + \frac{Lw}{c_p}. \tag{2.32}$$

(d) Isentropic condensation temperature T_c, defined as the temperature at which saturation is reached when moist air is cooled adiabatically with w held constant. This temperature is most readily understood with the help of a thermodynamic chart (Fig. 2.2).

The air initially has coordinates (T, p) with mixing ratio w. It is cooled adiabatically until its adiabat intersects the vapor line defined by $w_s = w$. The pressure at this intersection is called the isentropic condensation pressure, and the temperature is T_c. An analytical approximation for T_c, which must be solved by iteration, is

$$T_c = B/\ln\left[\frac{A\varepsilon}{wp_0}\left(\frac{T_0}{T_c}\right)^{1/k}\right]. \qquad (2.33)$$

This equation is obtained by noting that $T_c = T_d(w, p_c)$ and by substituting (2.28) for T_d into the adiabatic equation (1.23), written in the form

$$\frac{T_c}{T_0} = \left(\frac{p_c}{p_0}\right)^k.$$

Actually, it is not obvious that condensation should occur when expansion continues beyond the saturation point. Experience shows that this does happen in the atmosphere, so that we speak interchangeably of the condensation point and the saturation point.

Pseudoadiabatic process*

If expansion continues after the isentropic condensation point is reached, condensation occurs and the released latent heat will tend to warm the air. As a result, the temperature will decrease with falling pressure at a slower rate after condensation than before. To calculate the dependence of T on p in this process, assumptions must be made about the condensed water. Does it stay with the air in the form of cloud droplets or does it precipitate out? At subfreezing temperatures is the condensate water or ice? The different alternatives are compared in standard texts on dynamic meteorology. It turns out that the final result—the dependence of T on p—is not significantly affected by the choice of assumptions. The simplest case is the so-called pseudoadiabatic process in which the condensate is assumed to be water which immediately precipitates. This is the simplest case because the heat content of the condensed material need not be considered when calculating tem-

---

*In a brief history of early developments in the theory of the saturated adiabatic process, McDonald (1963a) explained that Lord Kelvin in 1862 gave the first correct description of the process; that Heinrich Hertz constructed an adiabatic diagram in 1884 that was the prototype of all subsequent meteorological thermodynamic charts; and that Wilhelm von Bezold in 1888 formulated the theory and equations for the pseudoadiabatic process.

perature changes of the air. Also, the question at what temperature sublimation becomes important is avoided.

Consider a sample of saturated air consisting of one kilogram of dry air and w_s kilograms of water vapor. Let its pressure change by amount dp. The temperature will change by dT and a corresponding change dw_s in vapor content will occur. The equation relating these incremental changes is, to good approximation,

$$\frac{dT}{T} = k\,\frac{dp}{p} - \frac{L}{Tc_p}\,dw_s, \tag{2.34}$$

a mathematical description of the pseudoadiabatic process. This formula is the basis of "pseudoadiabats" on a thermodynamic chart.

It is straightforward to employ an analytical approximation for $e_s(T)$, such as (2.12), and to express w_s in terms of differentials of temperature and pressure. When this expression is substituted in (2.34), the terms in dT and dp can be collected to solve for dT/dp in a pseudoadiabatic process. This relationship can then be integrated numerically to give temperature as a function of pressure along a pseudoadiabat.

Figure 2.3 illustrates the pseudoadiabatic expansion process. In an adiabatic expansion, the temperature decreases along a dry adiabat until the isentropic condensation point P is reached. Continued expansion is accompanied by the release of latent heat and the temperature follows along a pseudoadiabat from P onwards.

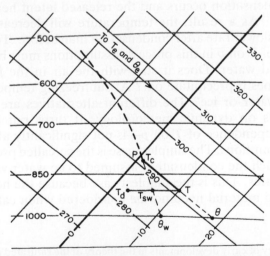

FIG. 2.3. Continued expansion of the air sample of Fig. 2.2 beyond point P, the isentropic condensation point. The dashed line is the pseudoadiabat through P. This diagram indicates the graphical determination of some of the important theoretical temperatures that characterize an air sample.

Some additional special temperatures that may be defined by referring to this illustration are:

(a) Adiabatic wet-bulb temperature T_{sw}, obtained by following the pseudoadiabat from P down to the original pressure. The result is within 0.5°C of the wet-bulb temperature defined by (2.30).

(b) Wet-bulb potential temperature θ_w, defined by the intersection of the pseudoadiabat through P with the isobar $p = 1000$ mb.

(c) Equivalent temperature T_e (adiabatic definition), obtained by following up the pseudoadiabat from P to very low pressure, thus condensing out all the water vapor, and then returning to the original pressure along a dry adiabat. This temperature follows from integrating (2.34) from initial temperature T to final temperature T_e, and may be shown to be approximately

$$T_e = T \exp \left[\frac{Lw_s}{c_p T_c} \right]. \tag{2.35}$$

(d) Equivalent potential temperature θ_e, defined as the temperature a parcel of air would have if taken from its equivalent temperature to a pressure of 1000 mb in a dry adiabatic process. A semi-empirical formula for θ_e, accurate to within 0.5 K, is

$$\theta_e = \theta \exp (2675w/T_c). \tag{2.36}$$

There is a one-to-one relationship between θ_e and θ_w. Both are determined by the process pseudoadiabat that characterizes the air sample; both are conservative in dry adiabatic or pseudoadiabatic processes. Highly accurate empirical formulas for these quantities are given by Bolton (1980).

The disposition of T, T_c, T_d, and T_{sw} about the perimeter of a triangle is called Normand's rule, after the British meteorologist Sir Charles Normand. It provides a helpful method of determining some of the thermodynamic properties of an air sample graphically.

Figure 2.4 is a tephigram including pseudoadiabats and vapor lines. It may be used for approximate graphical calculations.

Adiabatic liquid water content

In the pseudoadiabatic process the condensed phase is disregarded in calculating the changes in temperature of the air. The amount of this condensed material may be determined as follows. Suppose the air at the isentropic condensation point consists of 1 kg of dry air and w_s kg of water vapor. As this air expands pseudoadiabatically, the saturation mixing ratio decreases by the amount dw_s. If the air is to remain saturated, the

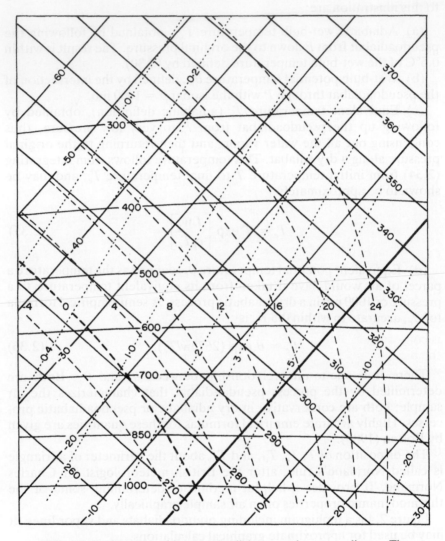

FIG. 2.4. Tephigram with pseudoadiabats and vapor lines. The pseudoadiabats are lines of constant θ_w and are labeled in degrees C at a position on the chart between 500 and 600 mb. The vapor lines are lines of constant w_s and are labeled in g/kg.

same amount of water must be condensed. Denoting this amount by $d\chi$, we have

$$d\chi = -dw_s \qquad (2.37)$$

as the relation between the saturation mixing ratio and the amount of condensed material. We call $d\chi$ the increase in adiabatic liquid water mixing ratio. χ increases as expansion continues, and is given at any time by $\chi = \Delta w_s$, where Δw_s is the decrease in w_s from its value at the condensation point. Note that the mass of liquid water per unit mass of moist air is $\chi/(1 + w_s)$. The liquid water density, the mass per unit volume of air, equals $\varrho\chi$, where ϱ is the air density. All of these quantities are measures of the adiabatic liquid water content of air produced by pseudoadiabatic expansion beyond the condensation point.

Reversible saturated adiabatic process

In this process, the condensed water remains with the air and its heat capacity is taken into account. From (2.37) it follows that the total water mixing ratio Q, defined by

$$Q = w_s + \chi, \qquad (2.38)$$

is conserved in a saturated air parcel. Moreover, on the assumption that the parcel is a closed thermodynamic system, the changes in temperature will be reversible, adiabatic, and isentropic.

The specific entropy of the cloudy air is given by

$$\phi = \phi_d + w_s\phi_v + \chi\phi_w, \qquad (2.39)$$

where ϕ_d, ϕ_v, and ϕ_w are respectively the specific entropies of the dry air, the vapor, and the liquid. From (2.4), we note that $\phi_v = \phi_w + (L/T)$, so that (2.39) may be written

$$\phi = \phi_d + \phi_w Q + \frac{L}{T}w_s. \qquad (2.40)$$

Therefore, in an isentropic process,

$$d\phi = 0 = d\phi_d + d(\phi_w Q) + d(Lw_s/T). \qquad (2.41)$$

But

$$d\phi_d = c_p d(\ln T) - R'd(\ln p_d)$$

and

$$d\phi_w = c_w d(\ln T),$$

where c_w here stands for the specific heat of liquid water, so that

$$(c_p + Qc_w)d(\ln T) - R'd(\ln p_d) + d(Lw_s/T) = 0. \qquad (2.42)$$

The main difference between (2.42) and (2.34) is that the latent heat absorbed by the water substance is accounted for in a reversible saturated adiabatic process. For most applications, the two equations give very similar results for the change of temperature with pressure.

Integrating (2.42) gives the result

$$\frac{T}{p_d^{R'/(c_p + c_w Q)}} \exp\left[\frac{w_s L}{T(c_p + c_w Q)}\right] = \text{constant.}$$

By analogy with the definition of potential temperature, we can now define the wet equivalent potential temperature θ_q as

$$\theta_q(\text{sat.}) = T\left(\frac{100\text{ kPa}}{p_d}\right)^{R'/(c_p + c_w Q)} \exp\left[\frac{w_s L}{T(c_p + c_w Q)}\right]. \qquad (2.43)$$

θ_q is a conservative quantity in a reversible saturated adiabatic process.

If the air parcel is not saturated, it can be shown that

$$\theta_q(\text{unsat.}) = \varkappa T\left(\frac{100\text{ kPa}}{p_d}\right)^{(R' + R_v Q)/(c_p + c_{pv} Q)} \qquad (2.44)$$

is conserved along a dry adiabat. The constant $\varkappa$ can be evaluated by requiring that $\theta_q(\text{sat.}) = \theta_q(\text{unsat.})$ at the pressure level where condensation occurs.

Problems

2.1. Prove that pure water vapor can reach saturation in the following processes:

(a) isothermal compression;
(b) adiabatic expansion.

2.2. A sample of moist air has a temperature of $-5°C$, a pressure of 80 kPa, and a relative humidity of 65%. Solve for the following properties of the sample by calculations, using approximations where convenient. Confirm the answers with the tephigram wherever possible:

(a) potential temperature;
(b) mixing ratio;
(c) dew point;
(d) isentropic condensation temperature;
(e) wet-bulb temperature;
(f) wet-bulb potential temperature;
(g) equivalent temperature;
(h) virtual temperature;
(i) density.

2.3. Household humidifiers work by evaporating water into the air of a confined space and raising its relative humidity. A large room with a volume of 100 m³ contains air at 23°C with a relative humidity of 15%. Compute the amount of water that must be evaporated to raise the relative humidity to 65%. Assume an isobaric process at 100 kPa in which the heat required to evaporate the water is supplied by the air. Confirm your answer with the tephigram.

2.4. A pseudoadiabatic process is not a polytropic process but it may be approximated as such over a limited range of pressure and temperature. Show that in a pseudoadiabatic process at 270 K and 80 kPa, pressure and specific volume are related by $p\alpha^n = $ const., where n has the value 1.23. Show further that n deviates only 5% from this value as the temperature changes by ± 10 K about 270 K and the pressure changes by ± 10 kPa about 80 kPa.

2.5. Using (2.14) for the dependence of the latent heat of vaporization on temperature, integrate (2.10) obtaining an expression for $e_s(T)$. (This result is sometimes called the Magnus equation or Kiefer's formula.) Over the range $-30°C \le T \le 30°C$, compare $e_s(T)$ from this equation with the simpler approximation (2.12), with Bolton's empirical formula (2.17), and with the data in Table 2.1.

3

Parcel Buoyancy and Atmospheric Stability

Hydrostatic equilibrium

Air is said to be in hydrostatic equilibrium when it experiences no net force in the vertical direction. Then the vertical pressure gradient force on the air exactly balances the force of gravity (Fig. 3.1), and

$$\frac{\partial p}{\partial z} = -\varrho g, \tag{3.1}$$

the well-known hydrostatic equation. Integration of (3.1) shows that the hydrostatic pressure at any level in the atmosphere is equal to the weight of an air column with unit cross-sectional area extending upwards from that level.

Substituting for ϱ from the equation of state gives

$$\frac{dp}{p} = -\frac{g}{R'T_v} dz. \tag{3.2}$$

Integrating this equation, we get

$$p = p_0 \exp\left[-\frac{g}{R'\overline{T}_v}(z - z_0)\right], \tag{3.3}$$

where p is pressure at height z, and $\overline{T}_v$ is the mean virtual temperature

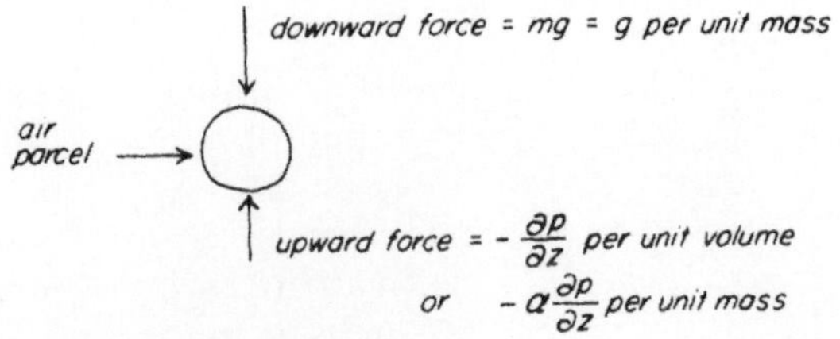

downward force $= mg = g$ per unit mass

air parcel $\longrightarrow$

upward force $= -\dfrac{\partial p}{\partial z}$ per unit volume

or $\quad -\alpha\dfrac{\partial p}{\partial z}$ per unit mass

FIG. 3.1. A parcel of air in hydrostatic equilibrium.

over the pressure interval from p to p_0, given by

$$\overline{T}_v = \frac{\int_{\ln p_0}^{\ln p} T_v d(\ln p)}{\ln p - \ln p_0}. \tag{3.4}$$

Dry adiabatic lapse rate

For dry air undergoing an adiabatic change of pressure,

$$c_p dT = \frac{R' T}{p} dp.$$

Thus for dry air ascending and expanding

$$\frac{dT}{dz} = \frac{R'}{c_p} \frac{T}{p} \frac{dp}{dz}. \tag{3.5}$$

The pressure in an unconfined sample (parcel) of air will immediately adjust to the ambient pressure, so

$$\frac{dp}{dz} = \frac{\partial p}{\partial z} = -\varrho' g, \tag{3.6}$$

where ϱ' denotes ambient density:

$$\varrho' = \frac{p}{R' T'} \tag{3.7}$$

with T' the ambient temperature.

Combining these equations shows that

$$\frac{dT}{dz} = -\frac{g}{c_p} \frac{T}{T'}.$$

Because the temperature of the parcel is not too different from ambient temperature, $T/T' \approx 1$ and the result simplifies to

$$\frac{dT}{dz} = -\frac{g}{c_p} \equiv -\Gamma, \tag{3.8}$$

where $\Gamma = g/c_p = 0.98°\text{C}/100$ m $\approx 10^{-2}$ K/m denotes the *dry adiabatic lapse rate*. This is the rate at which temperature falls off with height in the process of dry adiabatic ascent. It can be shown that the adiabatic lapse rate for moist (but unsaturated) air is equal to Γ to a close approximation.

The assumption that the pressure in an unconfined parcel of air adjusts immediately to the hydrostatic pressure is valid for large-scale vertical air motion but not necessarily for cumulus convection. Parcel theory usually neglects the pressure deviations from their hydrostatic values, but the dynamic effects of these deviations can be important (Yau, 1979).

Buoyant force on a parcel of air

Consider a parcel of dry air with volume V having temperature T and density ϱ. It displaces an equal volume of ambient air having temperature T' and density ϱ'. The downward force on the parcel is equal to $\varrho g V$. The downward force on the air displaced is equal to $\varrho' g V$. The upward force is the same for parcel and displaced air, $-V(\partial p/\partial z)$. Hence the net buoyant force (upward) is $Vg(\varrho' - \varrho)$. Therefore, the buoyant force per unit mass is

$$F_B = g\left(\frac{\varrho' - \varrho}{\varrho}\right) = g\left(\frac{T - T'}{T'}\right). \tag{3.9}$$

If this is the only force acting on a parcel, its equation of motion is

$$\frac{d^2z}{dt^2} = F_B = g\left(\frac{T - T'}{T'}\right). \tag{3.10}$$

As expected, this force is positive when the parcel is warmer than ambient air, negative when the parcel is cooler than ambient. For moist air, (3.9) may be generalized by merely replacing the temperatures with virtual temperatures.

Stability criteria for dry air

One of the uses of the dry adiabatic lapse rate is in assessing the stability of atmospheric layers with respect to the vertical displacement of a parcel. If after a small vertical displacement the parcel at a given level is subject to a restoring force which tends to accelerate it toward its original position, the atmosphere at that level is said to be stable. If after displacement the parcel is subject to a force in the direction of the displacement, the atmosphere is said to be unstable. The stability condition depends on the ambient lapse rate, that is, the decrease of temperature with height at the level of the test parcel.

Consider a parcel of air with the ambient temperature T initially. If it is lifted adiabatically a small distance Δz it cools by the amount $\Gamma \Delta z$ and its temperature is reduced to $T - \Gamma \Delta z$. Let us denote the ambient lapse rate by γ, that is,

$$-\left(\frac{\partial T}{\partial z}\right) = \gamma,$$

which is not to be confused with a *process* lapse rate. At height Δz above the initial position of the parcel, the ambient temperature is $T - \gamma \Delta z$. The excess temperature of parcel over ambient air is therefore $\Delta z(\gamma - \Gamma)$.

When this quantity is positive the parcel is warmer than its surroundings and, by (3.9), is accelerated upwards. Consequently the air is unstable whenever $\gamma - \Gamma > 0$. Conversely the parcel is subjected to a restoring force (downward) whenever $\gamma - \Gamma < 0$. For the special case $\gamma = \Gamma$, the displaced parcel experiences zero buoyancy force. The stability criteria for dry air may thus be summarized

$$\gamma < \Gamma \quad \text{STABLE}$$

$$\gamma = \Gamma \quad \text{NEUTRAL}$$

$$\gamma > \Gamma \quad \text{UNSTABLE}$$

These criteria may alternatively be expressed in terms of potential temperature. From the defining equation (1.24), the differentials of T, θ, and p are related by

$$dT = \frac{T}{\theta}\, d\theta + \frac{kT}{p}\, dp. \tag{3.11}$$

Also, we may take partial derivatives with respective to height and obtain

$$\frac{1}{\theta}\frac{\partial\theta}{\partial z} = \frac{1}{T}\frac{\partial T}{\partial z} - \frac{k}{p}\frac{\partial p}{\partial z} = \frac{1}{T}(\Gamma - \gamma), \tag{3.12}$$

where we have employed (3.1), (3.8), and the equation of state. Equivalently, therefore, the stability conditions may be described by

$$\frac{\partial\theta}{\partial z} > 0 \quad \text{STABLE}$$

$$\frac{\partial\theta}{\partial z} = 0 \quad \text{NEUTRAL}$$

$$\frac{\partial\theta}{\partial z} < 0 \quad \text{UNSTABLE}$$

The stability conditions can also be obtained directly from (3.10). By expanding $T(z)$ and $T'(z)$ as Maclaurin's series in z, and neglecting the higher order terms, we find that (3.10) reduces to

$$\frac{d^2z}{dt^2} = -\frac{g}{T}(\Gamma - \gamma)z. \tag{3.13}$$

From (3.12), this result may be written

$$\frac{d^2z}{dt^2} = -\frac{g}{\theta}\left(\frac{\partial\theta}{\partial z}\right)z \equiv -N^2z, \tag{3.14}$$

where $N = \sqrt{(g/\theta)(\partial\theta/\partial z)}$ has units of s^{-1} and is called the Brunt–Väisälä frequency.

For $N^2 = 0$, the displaced parcel is in neutral equilibrium and there is no restoring force. For $N^2 > 0$, the equilibrium is stable and the parcel undergoes oscillatory motion about its initial point. A typical value of N in the atmosphere is about $1.2 \times 10^{-2}\,\text{s}^{-1}$, so that the period of oscillation $\tau = 2\pi/N$ is about 8 min. For $N^2 < 0$, the equilibrium is unstable and the displacement grows exponentially. Obviously, the stability conditions based on N^2 are the same as those given by $\partial\theta/\partial z$.

The pseudoadiabatic lapse rate

Differentiating (2.34) with respect to height gives for the pseudo-adiabatic process

$$\frac{dT}{dz} = \frac{kT}{p}\frac{dp}{dz} - \frac{L}{c_p}\frac{dw_s}{dz}. \tag{3.15}$$

By employing the hydrostatic equation and the Clausius–Clapeyron equation, this can be reduced to an expression for the pseudoadiabatic (or saturated adiabatic) lapse rate:

$$\Gamma_s \equiv -\frac{dT}{dz} = \Gamma\frac{\left[1 + \dfrac{Lw_s}{R'T}\right]}{\left[1 + \dfrac{L^2\varepsilon w_s}{R'c_pT^2}\right]}. \tag{3.16}$$

It can be seen from (3.16) that $\Gamma_s < \Gamma$ whenever $L\varepsilon > c_pT$. Owing to the high value of L for water, this inequality is always satisfied in the atmosphere.

Stability criteria for moist air

When a saturated parcel is displaced upwards its temperature will decrease at the pseudoadiabatic rate. If the ambient lapse rate is greater than pseudoadiabatic, the displaced parcel will find itself warmer than its surroundings and will be accelerated in the direction of the displacement. Such air is unstable with respect to pseudoadiabatic parcel displacement. Allowing for the possibility of condensation on ascent leads to five possible states of stability for moist air:

$$\gamma < \Gamma_s \qquad \text{ABSOLUTELY STABLE}$$

$$\gamma = \Gamma_s \qquad \text{SATURATED NEUTRAL}$$

$$\Gamma_s < \gamma < \Gamma \qquad \text{CONDITIONALLY UNSTABLE}$$

$$\gamma = \Gamma \qquad \text{DRY NEUTRAL}$$

$$\gamma > \Gamma \qquad \text{ABSOLUTELY UNSTABLE}$$

Convective instability

The weight of a column of air with unit cross sectional area which extends from pressure level p_1 up to p_2 is equal to $(p_1 - p_2)$. We consider vertical displacements of this column in which its weight remains constant. (Since g is constant to good approximation, this is equivalent to vertical displacements with mass remaining constant.) Under this condition, $\Delta p = p_1 - p_2$ is constant.

From the hydrostatic equation, the height of the column and its pressure-thickness are related by $\Delta p = g \, \overline{\varrho} \Delta z$ with $\overline{\varrho}$ the mean density of air in the column. Since ϱ decreases with height, it follows that the lifting process considered here results in changes in Δz: stretching accompanies lifting of the column and contraction accompanies lowering. Usually the stability of the air will be affected by this process.

Consider first the lifting of dry air. Before displacement the stability of the air is measured by $\partial\theta/\partial z$ according to the criteria given previously. Thus over a small height interval δz the potential temperature varies by an amount equal to $\delta\theta = (\partial\theta/\partial z)\delta z$. For this incremental layer, $\delta\theta$ remains constant in adiabatic displacements. However, when the layer is lifted subject to the constraint described δz will increase in consequence of the stretching and it follows that $(\partial\theta/\partial z)$ must decrease. On the other hand $(\partial\theta/\partial z)$ must increase if the column of air is lowered. An exception to this is air having neutral stability, for which $(\partial\theta/\partial z) = 0$ before and after displacement.

These results mean that lifting does not affect the stability of an initially neutral layer. An initially unstable layer becomes less unstable; an initially stable layer becomes less stable. In short, lifting makes the lapse rate tend toward the dry adiabatic. Lowering a layer on the other hand makes its lapse rate depart further from adiabatic.

This effect may be readily illustrated on a thermodynamic chart. Shown in Fig. 3.2 is a layer of 100 mb thickness before and after lifting for the three possible stability conditions. It is noteworthy that air initially stable remains so after being lifted.

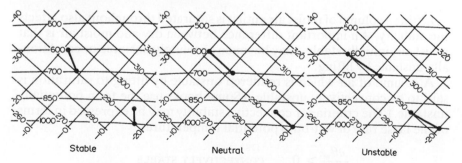

Stable   Neutral   Unstable

FIG. 3.2. Effect of lifting on stability of a layer of dry air.

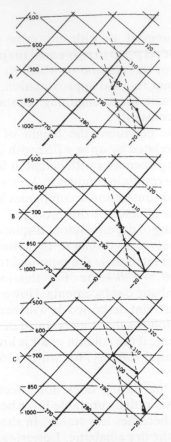

FIG. 3.3. Illustration of criterion for convective instability. In each case the layer between 900 and 1000 mb is lifted to 700 mb and condensation occurs throughout the layer. In case A, $\partial\theta_w/\partial z > 0$ and lifting to condensation stabilizes the layer. Case B has neutral convective stability with $\partial\theta_w/\partial z = 0$. Case C, $\partial\theta_w/\partial z < 0$, is convectively unstable.

Lifting a column of moist air until it is saturated throughout also affects the stability. An important difference between dry and moist air is that moist air, initially stable, may be made absolutely unstable or conditionally unstable by lifting.

A column of air which is rendered unstable by lifting to saturation is said to be convectively unstable. (Some books use the term potentially unstable.) The criteria for convective stability may be expressed in terms of the lapse rate of wet-bulb potential temperature.

$$\frac{\partial\theta_w}{\partial z} > 0 \qquad \text{CONVECTIVELY STABLE}$$

$$\frac{\partial \theta_w}{\partial z} = 0 \qquad \text{CONVECTIVELY NEUTRAL}$$

$$\frac{\partial \theta_w}{\partial z} < 0 \qquad \text{CONVECTIVELY UNSTABLE}$$

These criteria are also best understood with reference to a tephigram (Fig. 3.3).

Convective instability has to do with the lifting of layers and should not be confused with conditional instability, which applies to an undisplaced layer. A layer that is conditionally unstable need not be convectively unstable; nor is a convectively unstable layer necessarily conditionally unstable.

Horizontal restoring forces

So far, our discussion of stability conditions has been limited to the vertical displacement of an air parcel or a layer of air. In the atmosphere, instability can also occur when air is displaced in a slantwise direction. The requirements for slantwise instability can be understood by including horizontal restoring forces in the stability analysis.

The major horizontal forces in the atmosphere are the Coriolis force and the horizontal pressure gradient force. The nature of the Coriolis force can be illustrated by considering an object moving from point C to point B on a turntable rotating counterclockwise as depicted in Fig. 3.4. If the object follows a true straight line in space, the projection of its trajectory on the turntable will actually be the curve from C to A. To an observer on the rotating system, it would appear that a horizontal force

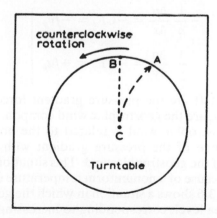

FIG. 3.4. Demonstration of the effect of the Coriolis force in the northern hemisphere. (From Panofsky, 1981.)

has deflected the object toward the right. The magnitude of this force, called the Coriolis force, is equal to twice the product of the local rate of rotation and the speed of the object. Applying this result to the movement of an air parcel in the northern hemisphere, we see that the Coriolis force acts in a direction normal to the trajectory of the parcel and tends to deflect it toward the right. Let x denote the eastward direction and y the northward direction. Then the components of the deflecting force in these directions are given by

$$\left. \begin{array}{l} CF_x = 2\Omega \sin \phi v = fv \\ CF_y = -2\Omega \sin \phi u = -fu \end{array} \right\} \quad (3.17)$$

where Ω is the angular velocity of the earth's rotation (7.29×10^{-5} rad/s), ϕ is the latitude, and u and v are the x and y components of the horizontal wind. $\Omega \sin \phi$ is the local rate of rotation at latitude ϕ. $f = 2\Omega \sin \phi$ is known as the Coriolis parameter.

Geostrophic wind and geostrophic wind shear

The variation of pressure in the horizontal creates horizontal pressure gradient forces, which accelerate an air parcel from high to low pressure. However, as soon as the parcel begins to move, the Coriolis force acts to deflect it toward the right in the northern hemisphere. The magnitude of the force increases as the parcel picks up speed, and eventually a state of balance can be achieved in which the Coriolis force exactly balances the horizontal pressure gradient force. In this state, the air motion and hence the wind is parallel to the isobars. The forces are at right angles to the wind, as shown in Fig. 3.5. The balanced flow, called the geostrophic wind, is described by the relations

$$\left. \begin{array}{l} -\dfrac{1}{\varrho} \dfrac{\partial p}{\partial x} = -CF_x = -fv_g \\[3mm] -\dfrac{1}{\varrho} \dfrac{\partial p}{\partial y} = -CF_y = +fu_g \end{array} \right\} . \quad (3.18)$$

The terms on the left are the pressure gradient forces in the x and y directions. u_g and v_g are the geostrophic wind components.

Because the geostrophic wind is related to the horizontal pressure gradient, any change of the pressure gradient with height implies a vertical variation of the geostrophic wind. This situation often prevails in the atmosphere because of a nonuniform temperature distribution in the horizontal. Figure 3.6 shows a situation in which the geostrophic wind is from the west at sea level, corresponding to increasing surface pressure from north to south. If the air is colder to the north, as shown, the hydrostatic equation then requires that the pressure decrease more

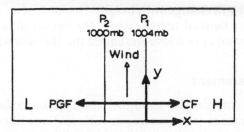

FIG. 3.5. Geostrophic wind in the northern hemisphere, a balance between the pressure gradient force (PGF) and the Coriolis force (CF). (From Panofsky, 1981.)

rapidly with height in the north. The horizontal pressure gradient force at 3 km is therefore stronger than at sea level and the speed of the west wind increases with height.

The variation of the geostrophic wind with height, or the geostrophic wind shear, can be obtained by differentiating (3.18) with respect to z and making use of the hydrostatic equation and the definition of potential temperature. To a good approximation, the result is

$$\left.\begin{array}{l} -f\dfrac{\partial v_g}{\partial z} = \dfrac{\partial}{\partial z}\left(-\dfrac{1}{\varrho}\dfrac{\partial p}{\partial x}\right) = -\dfrac{g}{\theta}\dfrac{\partial \theta}{\partial x} \\[2ex] f\dfrac{\partial u_g}{\partial z} = \dfrac{\partial}{\partial z}\left(-\dfrac{1}{\varrho}\dfrac{\partial p}{\partial y}\right) = -\dfrac{g}{\theta}\dfrac{\partial \theta}{\partial y} \end{array}\right\}. \tag{3.19}$$

This shows that the geostrophic wind shear is related to the horizontal gradient of potential temperature. If the potential temperature is uniform in the horizontal, which means that the surfaces on which θ is constant are horizontal, the geostrophic wind does not vary with height. On the other hand, if the surfaces of constant θ are not horizontal, the

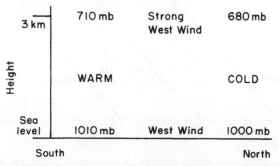

FIG. 3.6. Change of geostrophic wind with height, caused by a horizontal temperature gradient.

geostrophic wind will change with height. Because of its dependence on the horizontal gradient of temperature, the vector difference between the geostrophic wind at two levels is called the thermal wind.

Slantwise displacement

In Fig. 3.7 we consider the situation in which the surfaces of constant potential temperature (the isentropic surfaces) are tilted as indicated. Suppose a parcel of air at point A is in equilibrium with the environment. That is, the parcel has the same temperature, potential temperature, pressure, and velocity as its environment. Next we suppose that the parcel is displaced slantwise to position B. If condensation does not occur the potential temperature is conserved and the temperature of the parcel at B is

$$T + \left(\frac{dT}{dp}\right)dp = T + \frac{kT}{p}\left(\frac{\partial p}{\partial y}\,\delta y + \frac{\partial p}{\partial z}\,\delta z\right).$$

The ambient temperature at B is given by

$$T + \frac{\partial T}{\partial y}\,\delta y + \frac{\partial T}{\partial z}\,\delta z = T + \left(\frac{T}{\theta}\frac{\partial \theta}{\partial y} + \frac{kT}{p}\frac{\partial p}{\partial y}\right)\delta y$$
$$+ \left(\frac{T}{\theta}\frac{\partial \theta}{\partial z} + \frac{kT}{p}\frac{\partial p}{\partial z}\right)\delta z,$$

so that the excess temperature of the displaced parcel over the ambient air is

$$-T\left(\frac{1}{\theta}\frac{\partial \theta}{\partial z}\,\delta z + \frac{1}{\theta}\frac{\partial \theta}{\partial y}\,\delta y\right),$$

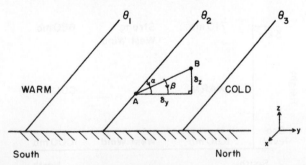

FIG. 3.7. Assessment of parcel stability for slantwise displacement from an equilibrium condition.

and the expression for the buoyancy force on the displaced parcel is

$$F_B = -g\left(\frac{1}{\theta}\frac{\partial\theta}{\partial z}\,\delta z + \frac{1}{\theta}\frac{\partial\theta}{\partial y}\,\delta y\right). \tag{3.20}$$

In addition to the buoyancy force, the parcel is subject to a horizontal restoring force because the Coriolis and horizontal pressure forces acting on the parcel are out of balance at the new position. If the parcel moves from A to B in time δt, the Coriolis force changes the x-component of its velocity by an amount $fv\delta t = f\delta y$. At its new position, the Coriolis force on the parcel is therefore increased by the amount $f^2\delta y$. The tilt of the isentropic surfaces indicates that the horizontal pressure gradient force must vary with height. From (3.19), the change of the pressure force from A to B is given by

$$\frac{\partial}{\partial y}\left(-\frac{1}{\varrho}\frac{\partial p}{\partial y}\right)\delta y + \frac{\partial}{\partial z}\left(-\frac{1}{\varrho}\frac{\partial p}{\partial y}\right)\delta z = f\frac{\partial u_g}{\partial y}\,\delta y + f\frac{\partial u_g}{\partial z}\,\delta z.$$

Because the parcel is in equilibrium at A, the net horizontal restoring force F_H at B is given by the difference between the incremental changes in the Coriolis force and the horizontal pressure gradient force. That is,

$$F_H = f\left[\frac{\partial u_g}{\partial z}\,\delta z - \left(f - \frac{\partial u_g}{\partial y}\right)\delta y\right]. \tag{3.21}$$

The equation of motion of the parcel along its direction of displacement, with distance denoted by Δ, is therefore

$$\frac{d^2\Delta}{dt^2} = F_B\sin\beta + F_H\cos\beta$$

$$= -g\left[\frac{1}{\theta}\frac{\partial\theta}{\partial z}\,\delta z + \frac{1}{\theta}\frac{\partial\theta}{\partial y}\,\delta y\right]\sin\beta$$

$$+ f\left[\frac{\partial u_g}{\partial z}\,\delta z - \left(f - \frac{\partial u_g}{\partial y}\right)\delta y\right]\cos\beta. \tag{3.22}$$

This is a generalized equation for parcel displacement, which reduces to (3.14) for vertical displacement by setting $\delta y = 0$ and $\beta = 90°$.

Symmetric instability

A type of slantwise instability, called symmetric instability, can exist for an air parcel that is displaced along an isentropic surface. The buoyancy force then vanishes and the restoring force is given only by F_H. In this situation, the equation analogous to (3.22) becomes

$$\frac{d^2\Delta}{dt^2} = f\delta y \cos \beta \frac{\partial u_g}{\partial z} \left[\frac{\delta z}{\delta y} - \frac{\left(f - \frac{\partial u_g}{\partial y} \right)}{\left(\frac{\partial u_g}{\partial z} \right)} \right]. \tag{3.23}$$

If $\partial u_g/\partial z > 0$ and the sign of the factor in square brackets is negative, the displaced parcel will be accelerated towards its initial position, and we describe the initial equilibrium as stable. If this factor is positive, the equilibrium is unstable. The first term in this factor, $\delta z/\delta y$, is just the slope of the isentropic surface. The second term is related to the vorticity of the parcel.

Vorticity is a measure of the local rotational characteristics of a fluid. If the fluid is rotating as a solid body, its vorticity is twice the angular velocity. The vorticity of the atmosphere due to the rotation of the earth equals f. This is called the planetary vorticity and is directed vertically. The absolute vorticity of an air sample is the planetary vorticity plus the vorticity of the air relative to the earth. The relative vorticity of the westerly component of the geostrophic wind, u_g, has two components. Its y component is $\partial u_g/\partial z$ and its z component is $-\partial u_g/\partial y$, so that the z-component of the absolute vorticity is $f - (\partial u_g/\partial y)$. Therefore, in (3.23), the second term in brackets is the ratio of the horizontal to the vertical component of absolute vorticity, and may be interpreted as the slope of the absolute vorticity vector.

The stability conditions for symmetric instability may accordingly be described by

Slope of isentropic surface
> < Slope of absolute vorticity vector   STABLE

Slope of isentropic surface
> = Slope of absolute vorticity vector   NEUTRAL

Slope of isentropic surface
> > Slope of absolute vorticity vector   UNSTABLE

Bennetts and Hoskins (1979) and Emanuel (1979) analyzed this type of instability. They found that in a saturated, cloudy atmosphere, the stability conditions are similar to those for dry air with the exception that the slope of the isentropic surface is replaced by the slope of the surface of wet-bulb potential temperature. Bennetts and Sharp (1982) and Seltzer *et al.* (1985) showed that symmetric instability can be responsible for the mesoscale banded structure of precipitation associated with midlatitude frontal cyclones, as discussed in Chapter 12.

Baroclinic instability

Another type of slantwise instability can occur if only the generalized buoyancy force is included. The equation of motion for the air parcel is then

$$\frac{d^2\Delta}{dt^2} = -g\left(\frac{1}{\theta}\frac{\partial\theta}{\partial z}\right)\delta y \sin \beta \left[\frac{\delta z}{\delta y} - \left(-\frac{\frac{\partial\theta}{\partial y}}{\frac{\partial\theta}{\partial z}}\right) \right]. \qquad (3.24)$$

Because $\delta z/\delta y$ is now the slope of air parcel displacement and $-(\partial\theta/\partial y)/(\partial\theta/\partial z)$ is the slope of the isentropic surface, the stability criteria for a statically stable atmosphere ($\partial\theta/\partial z > 0$) become

Slope of isentropic surface < Slope of parcel displacement   STABLE

Slope of isentropic surface = Slope of parcel displacement   NEUTRAL

Slope of isentropic surface > Slope of parcel displacement   UNSTABLE

Charney (1947) and Eady (1949) were among the first to investigate this so-called baroclinic instability mechanism. The criteria for baroclinic instability are generally met in the atmosphere at midlatitudes and it is firmly established that this kind of instability is responsible for the formation of midlatitude cyclones and the associated widespread cloud and precipitation.

Geopotential

The geopotential $\psi(z)$ is defined as the potential energy of a unit mass at height z above a reference level, usually mean sea level. By this definition, $d\psi = gdz$ is the increase in geopotential for an incremental increase in altitude. Thus there is a close relation between the geopotential at level z and the geometric altitude. As a convenient artifice, the geopotential is measured in units of "geopotential meters", defined by

$$\text{gpm}(z) = \frac{\psi(z)}{g_0} = \frac{1}{g_0}\int_0^z gdz' = \frac{\overline{g}}{g_0}z,$$

where g_0 is a standard value of gravity, 9.8 m/s^2, and $\overline{g}$ is the average value of g between sea level and height z. Meteorological thermodynamic charts such as the tephigram usually have altitude scales, which are derived from the relation between the height in geopotential meters and the pressure in a standard atmosphere. The geopotential height, temperature, pressure, and density in the International Civil Aviation

Organization (ICAO) Standard Atmosphere are given as functions of geometric altitude in the Appendix.

Problems

3.1. Two model atmospheres that are often used in theoretical work are the homogeneous atmosphere, defined by

$$\varrho = \text{const.} = \varrho_0,$$

and the exponential atmosphere, defined by

$$\varrho = \varrho(z) = \varrho_0 \exp\left(-z/H\right),$$

where ϱ_0 is the density at the surface and H is called the scale height of the atmosphere. The top of the homogeneous atmosphere is defined as the altitude where its pressure falls to zero. Prove that the height of the top of the homogeneous atmosphere equals the scale height of the exponential atmosphere.

3.2. (a) Prove that the geopotential ψ and the specific enthalpy h of an air sample undergoing a dry adiabatic process are related by $h + \psi = \text{const.}$
(b) Show that the geopotential at pressure level p of an atmosphere in hydrostatic equilibrium is given by

$$\psi(p) = R'\overline{T}_v \ln\left(p_0/p\right),$$

where $\psi(p_0) = 0$.

3.3. By taking account of the variation of gravity with height, show that the altitude in geopotential meters is related to geometric altitude, to good approximation, by

$$\text{gpm} = z - az^2.$$

Determine the numerical value of a, and solve for gpm at 1, 10, and 50 km.

3.4. In an unstable layer of air over the ground the temperature decreases linearly with height at a rate of 3°C/100 m. A parcel of air at the bottom of this layer with a temperature of 280 K is given an initial upward velocity of 0.5 m/s. Assuming that the parcel ascends dry adiabatically with no aerodynamic resistance, show that after 1 min has elapsed it is approximately 45 m above the ground and ascending at 1.28 m/s.

3.5. When deflated, a hot-air balloon, consisting of air bag, gondola, supporting cables, fuel, and burner, has a weight of 1100 N. The balloon has a volume of 1000 m³ when inflated. Estimate the temperature to which the air in the balloon must be heated for it to float at a steady altitude of 2 km while carrying passengers with a combined weight of 1500 N. Assume a standard atmosphere, and assume that the pressure inside the balloon equals that of the ambient air.

3.6. The boiling point of a liquid is the temperature at which its vapor pressure equals the atmospheric pressure. Derive an expression for the change in the boiling point of water with altitude. Show that for typical sea level conditions the boiling point falls with altitude at a rate of about 3°C per km.

3.7. The mass of water vapor in a vertical column of unit cross sectional area extending from the surface to altitude z is called the columnar vapor. Sometimes called "precipitable water", this quantity has units of kg/m² in the SI system or g/cm² in the CGS system. Columnar vapor is sometimes defined as the volume equivalent of this mass, in which case the units are m³/m² = m in the SI system or cm³/cm² = cm in CGS.

Derive an expression for the columnar vapor between the surface and altitude z in terms of the mixing ratio. Evaluate this expression assuming that (1) the mixing ratio decreases exponentially with height with a scale height H_w and (2) the density decreases exponentially with height with a scale height H. That is,

$$w(z) = w_0 \exp(-z/H_w)$$
$$\varrho(z) = \varrho_0 \exp(-z/H).$$

For $H = 8\,\text{km}$, $H_w = 4\,\text{km}$, and $\varrho_0 = 1\,\text{kg/m}^3$, show that the total columnar water vapor in cm is given in terms of the surface mixing ratio by

$$W = 2.67 \times 10^2 \, w_0.$$

3.8. Condensation of water can occur in updrafts because the saturation mixing ratio decreases in adiabatic ascent. This property of water can be attributed to the high value of the latent heat of condensation. It has long been speculated that there may be trace gases which, because of low values of L, would condense in downward moving air (Bohren, 1986). Show that the criterion that must be satisfied if vapor is to condense in downdrafts is

$$L < c_p T/\varepsilon.$$

4

Mixing and Convection

Mixing of air masses

(a) Isobaric mixing

Consider two masses of moist air at pressure p: the first with mass M_1, temperature T_1, and specific humidity q_1; the second with mass M_2, temperature T_2, and specific humidity q_2. Suppose these two samples are thoroughly mixed at constant pressure.

The specific humidity of the mixture is a mass-weighted mean of the individual specific humidities,

$$q = \frac{M_1}{M_1 + M_2} q_1 + \frac{M_2}{M_1 + M_2} q_2. \tag{4.1}$$

From (2.18) and (2.19) it follows that, to close approximation, the mixing ratio and the vapor pressure of the mixture are also weighted means,

$$w \approx \frac{M_1}{M_1 + M_2} w_1 + \frac{M_2}{M_1 + M_2} w_2 \tag{4.2}$$

$$e \approx \frac{M_1}{M_1 + M_2} e_1 + \frac{M_2}{M_1 + M_2} e_2. \tag{4.3}$$

If there is no net loss or gain of heat during mixing, the amount of heat lost by the warmer sample is equal to the amount of heat gained by the cooler sample. That is, if T denotes the temperature of the mixture,

$$M_1(c_p + w_1 c_{pv})(T_1 - T) = M_2(c_p + w_2 c_{pv})(T - T_2).$$

Neglecting the small contribution of water vapor to this balance,

$$T \approx \frac{M_1}{M_1 + M_2} T_1 + \frac{M_2}{M_1 + M_2} T_2, \tag{4.4}$$

showing that the temperature of the mixture is a weighted mean of the temperatures of the two samples.

This kind of mixing process is readily described with a hygrometric chart, which is a plot of e against T (see Fig. 4.1). Each of the two samples

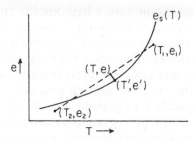

FIG. 4.1. Isobaric mixing of two air samples.

to be mixed is represented by a point on these coordinates, as illustrated. Equations (4.3) and (4.4) imply that the temperature and vapor pressure of the mixture correspond to some position on the straight line connecting these points. This position depends upon the ratio of M_1 to M_2. For example, if 4 kg of sample 1 are mixed with 3 kg of sample 2, the characteristic point of the mixture lies 4/7 the distance from point 2 up to point 1, as shown.

The solid line labeled e_s corresponds to the saturation vapor pressure, a function only of temperature according to the Clausius–Clapeyron equation (2.10). When two air masses are mixed under the conditions assumed here, the possibility exists that the mixture may be supersaturated, that is, have a relative humidity above 100%. In this event condensation occurs and a cloud forms in the mixture.[*] The hygrometric chart may also be used to illustrate this process.

In the example shown, the mixture is supersaturated, so condensation will occur. Condensation will proceed until the mixture is just saturated. This will not occur at temperature T, because during condensation the mixture will be heated by the latent heat of condensation. As the mixing ratio decreases by amount dw during condensation, the latent heating is given approximately by

$$dq = -L\,dw. \qquad (4.5)$$

Introducing (2.18) to convert from w to e, and assuming an isobaric process, we find

$$\frac{de}{dT} = -\frac{pc_p}{L\varepsilon}, \qquad (4.6)$$

---

[*]From the observation that breath is visible in cold weather, and using an argument similar to that depicted in Fig. 4.1, the Scottish geologist James Hutton in 1784 deduced the concave-upwards shape of the curve which was described by the Clausius–Clapeyron equation a century later. This and other interesting historical sidelights on cloud physics are described in a book by Middleton (1966).

which gives the slope of the line on a hygrometric chart describing the isobaric condensation process. The intersection of this line with the e_s curve defines a point with coordinates T', e', which corresponds to the mixture of air masses after condensation has occurred.

Two other quantities mix approximately linearly, as do w and e, and will be used later in the study of mixing processes in clouds. For an unmixed cloud parcel with a negligible amount of precipitation, the thermodynamic process can be regarded as saturated, reversible, and adiabatic. Therefore, as shown in Chapter 2, the total water mixing ratio Q and the wet equivalent potential temperature θ_q are conservative quantities.

If air parcels with total mixing ratios of Q_1 and Q_2 are mixed together, the mixture will have a total mixing ratio given by

$$Q \approx \left(\frac{M_1}{M_1 + M_2}\right)Q_1 + \left(\frac{M_2}{M_1 + M_2}\right)Q_2. \qquad (4.7)$$

Under normal atmospheric conditions, θ_q can also be shown to mix approximately linearly (see Problem 4.8). That is, the wet equivalent potential temperature of the mixture can be written as

$$\theta_q \approx \left(\frac{M_1}{M_1 + M_2}\right)\theta_{q1} + \left(\frac{M_2}{M_1 + M_2}\right)\theta_{q2}, \qquad (4.8)$$

where θ_{q1} and θ_{q2} denote the values of θ_q for the air parcels before mixing.

(b) Adiabatic mixing

Suppose the two samples of air start with different pressures and are mixed after being brought adiabatically to the same pressure. In this process of adiabatic mixing, the potential temperature of the mixture is a weighted mean of the potential temperatures of the two samples, in just the way that temperatures are related by (4.4). As before, the specific humidity of the mixture is given by (4.1).

Consequently, when a column of air is thoroughly mixed, the specific humidity will tend to a constant value throughout, given by

$$q_m = \frac{1}{M} \int_{z_1}^{z_2} \varrho q\, dz,$$

where $M = \int_{z_1}^{z_2} \varrho\, dz$ is the total mass of the column. This equation may be written in terms of the pressure-thickness of the column by introducing the hydrostatic equation, leading to

$$q_m = \frac{1}{\Delta p} \int_{p_2}^{p_1} q\, dp. \qquad (4.9)$$

Similar expressions apply to the mixing ratio and the vapor pressure of the mixture.

The potential temperature of the mixture tends to a constant value of

$$\theta_m = \frac{1}{\Delta p} \int_{p_2}^{p_1} \theta \, dp. \qquad (4.10)$$

With thorough mixing the temperature lapse rate in a vertical column thus approaches the dry adiabatic and the mixing ratio approaches a constant value; the limiting values of potential temperature and mixing ratio are averages with respect to pressure.

Convective condensation level

Vertical mixing within a column of air next to the ground occurs in consequence of solar heating of the surface. Heat is transferred by conduction from the surface to the air layer in contact with it, causing a strong lapse rate of temperature in the lowest layers of air. When the lapse rate becomes superadiabatic, any small disturbance will lead to vertical motions of elements of air in the layer, causing general mixing and overturning. The temperature profile in the mixing layer will tend toward the dry adiabatic and the mixing ratio will everywhere approach its average value with respect to pressure. If strong heating at the surface continues, this heat will be convected upwards, raising the potential temperature of the air throughout the layer. This process is indicated schematically in Fig. 4.2.

The heavy line indicates the initial temperature profile, with T_0 the surface temperature. Heating raises this temperature causing a super-adiabatic lapse rate temporarily and convection then tends to establish a dry adiabatic lapse rate with surface temperature T_1. Additional heating raises the temperature and increases the thickness of the mixing layer.

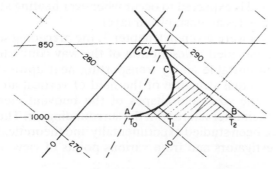

FIG. 4.2. Illustration of convective condensation level.

Eventually the surface temperature reaches T_2 and the total heat added by convection is proportional to the hatched area ABC. Meanwhile the mixing ratio has become approximately constant up to the altitude of point C and equal to the average in the layer. The dashed curve indicates the profile of T_d, assuming constant w. It can be seen that a small amount of additional heating will raise the top of the mixing layer to the point where the adiabat intersects the vapor line. Condensation is expected at this point, which is called the convective condensation level (CCL). This is the level where you would expect to find the bases of cumulus clouds. If the air is conditionally unstable above the CCL, ascent of rising convective elements would continue with temperature falling off at the pseudoadiabatic lapse rate. This schematic picture illustrates the principle of cloud formation by convection, but the real process may often be complicated by the fact that heating causes evaporation of water at the earth's surface, which will increase the mixing ratio, tending to lower the CCL.

Related to the CCL is the lifting condensation level (LCL), which is defined as the level where a parcel of air just reaches saturation if it is lifted dry adiabatically from the surface. The LCL depends only on the properties of the air at the surface and does not provide for any vertical mixing. In conditions under which cumulus clouds are observed to form, the LCL and CCL often agree closely with one another because the air below cloud base is well mixed.

Convection: elementary parcel theory

Convection refers to the vertical motions of elements of air. These motions can arise from buoyant or mechanical forces, and are the atmosphere's way of providing for efficient vertical transport of heat, mass, and momentum. Of special interest is buoyant convection, for this is the process leading to the formation of cumulus (or convective) clouds. Buoyant convection represents a conversion of potential energy to kinetic energy, and is expected to occur whenever heating at the surface or cooling aloft creates an unstable air layer.

Stability criteria were given in Chapter 3; the preceding section of the present chapter included a description of the way convection modifies the lapse rate in unstable air by transporting heat upwards. Not considered was the actual structure of the field of vertical air motion. Of importance are the sizes and shapes of the buoyant elements, their velocities, and their interaction with the surrounding air. These details of convection have been studied experimentally and theoretically by a large number of investigators and from various points of view over the past century.

Attention was first directed to the motion field established in incompressible fluids when heated uniformly at the bottom surface. This work was later extended by considering the effect of uniform or sheared motion of the fluid over the heated surface. More recent laboratory experiments with convection in fluids have shown striking similarities with cumulus clouds in the atmosphere and have led to theories of convection that account for some of the observed characteristics of clouds. Clouds themselves have been investigated by time lapse photography, instrumented airplanes, and radar to get an idea of the character of airflow in natural convection. Recently perfected techniques for remotely probing the atmosphere with acoustic or electromagnetic waves even permit the study of convective patterns in cloud-free air. Much of this work, though relevant to cloud physics, is beyond the scope of this book. However, some aspects of the physics of convection are presented in this and the following section, and later chapters treat the observation and numerical simulation of clouds. The interested reader is referred to Scorer's (1958) *Natural Aerodynamics* for a review of the dynamics of convection and to Gossard and Strauch (1983) for a summary of radar studies of clear-air convection.

The most elementary approach to finding the vertical velocity of a convective element is based on (3.10), which expresses the buoyant force on a parcel of air. It is assumed that the parcel—a buoyant element of air with size and shape unspecified—maintains its identity in thermodynamic processes; that it in no way disturbs or interacts with the environmental air; that it has uniform properties throughout; that its pressure instantly adjusts to the pressure of the surroundings. Following from (3.10), its equation of motion is

$$\frac{d^2z}{dt^2} = gB, \tag{4.11}$$

where

$$B = \frac{T - T'}{T'} \tag{4.12}$$

represents the buoyancy term. Introducing $U = dz/dt$ as the vertical velocity, (4.11) becomes

$$UdU = gBdz, \tag{4.13}$$

which may be integrated over height to give

$$U^2 = U_0^2 + 2g \int_{z_0}^{z} B(z)dz, \tag{4.14}$$

where U is the velocity at height z and U_0 is the velocity at z_0.

The integral $\int_{z_0}^{z} gBdz$ may be shown using the hydrostatic equation to be equal to $R' \int (T - T')d(\ln p)$. It represents the area on a thermodynamic chart bounded by the process curve of parcel temperature and the ambient temperature profile, from pressure $p(z_0)$ up to pressure $p(z)$. This area is proportional to the increase in kinetic energy of the buoyant parcel between z_0 and z. It is referred to as the "positive area" of the sounding.

The velocity predicted by (4.14) is likely to be too high for the following reasons:

1. Aerodynamic drag was neglected.
2. Mixing with ambient air was neglected.
3. Compensating downward motions of the surrounding air were neglected.
4. The weight of condensed water, some of which is carried along with the parcel, was neglected.

Therefore U in (4.14) may be interpreted as an expected upper limit for vertical velocity in buoyant convection.

Modification of the elementary theory

To describe more accurately the behavior of convective elements or "thermals", elementary parcel theory has been modified in different ways, in part as an attempt to overcome the shortcomings listed above. Some of these modifications are outlined in this section.

(a) The burden of condensed water

The buoyancy force per unit mass of air is gB for dry air, where B is given by (4.12). For moist air, the same expression holds if virtual temperature is used in place of temperature. If condensed water is present in the parcel, in the form of cloud droplets or precipitation, it exerts a downward force on the parcel equal to its weight. The buoyancy factor B then becomes

$$B = \frac{T}{T'} - (1 + \mu), \tag{4.15}$$

where μ denotes the "mixing ratio" of condensate, in terms of mass per unit mass of air. For adiabatic expansion with no mixing, and neglecting precipitation, $\mu = \chi$, the adiabatic liquid water content. The expression (4.15) assumes that there is no condensate in the ambient air around the thermal. This would not be so if the thermal were ascending through a cloud. A more general expression, allowing for this possibility, is

$$B = \frac{T}{T'} (1 + \mu') - (1 + \mu), \tag{4.16}$$

where μ' is the mixing ratio of condensate in the ambient air.

(b) Compensating downward motions

By the requirement of mass continuity, air must descend somewhere to replace the volume vacated by an upward-moving thermal. If the descending air is cloud-free, it will be warmed at the dry adiabatic rate. The air through which a thermal is ascending may thus have its temperature affected by adiabatic descent; the changes in temperature will then influence the buoyancy factor B.

The "slice-method" of stability analysis is designed to take into account this effect of ambient air descent. Attention is focused on a horizontal level through which thermals ascend and ambient air descends. The area occupied by thermals is denoted by A and the remainder, in which air is descending, is denoted by A'. The mass flux of upward-moving air through the level is ϱUA where U is the velocity of the thermals. The downward mass flux is $\varrho'A'U'$. It is assumed that the level is broad enough so that the upward and downward fluxes are equal. Therefore

$$\frac{A}{A'} = \frac{\varrho'U'}{\varrho U} \approx \frac{U'}{U}, \tag{4.17}$$

where it is assumed that $\varrho' \approx \varrho$.

It is further assumed that the ascending air is cooled at the pseudo-adiabatic rate while descending air is warmed at the dry adiabatic rate. Then, after a short time dt, the air arriving at the level from below will have a temperature given by $T_0 + \gamma U dt - \Gamma_s U dt$ where T_0 is the initial temperature at the level, Γ_s is the pseudoadiabatic lapse rate, and γ is the ambient lapse rate. The air arriving at the level from above will have temperature $T_0 - \gamma U' dt + \Gamma U' dt$. The situation is unstable when this temperature is less than the temperature of the thermal. Thus, for instability, we have

$$(\gamma - \Gamma_s)U > (\Gamma - \gamma)U'$$

or, from (4.17),

$$(\gamma - \Gamma_s)A' > (\Gamma - \gamma)A. \tag{4.18}$$

In the limit as $A \to 0$, this result reduces to the instability criterion of an elementary parcel, as given earlier.

Using the slice-method arguments shows further that the neutral case arises whenever

$$\frac{\gamma - \Gamma_s}{\Gamma - \gamma} = \frac{A}{A'}.$$

If $A/A' > 0$ (that is, if the thermals are not negligible in size), this equation can only be satisfied if $\gamma > \Gamma_s$. Thus the lapse rate must be steeper for instability when the compensating downward motions are taken into account than when they are neglected.

(c) Dilution by mixing

When a buoyant element ascends, some mixing is expected to take place through its boundaries. Since the ambient air is generally cooler and drier than the buoyant element, the mixing will tend both to reduce the buoyancy of the thermal and to lower its mixing ratio. This kind of mixing is called "entrainment", and theories have been developed to account for its thermodynamic effects.

In general, the effects of mixing can be analyzed by considering the heat exchange between the cloudy air and the entrained air. Consider a mass m of cloudy air, consisting of dry air, water vapor, and condensed water. If the mixing occurs with outside air entrained through the sides of the cloud, we can suppose that a mass dm of outside air is entrained as the cloud ascends through height dz. Let the cloudy air have temperature T, the ambient air T'. The heat required to warm the entrained air is then

$$dQ_1 = c_p(T - T')dm,$$

where the heat contents of the vapor and the condensate are assumed negligible compared to that of the dry air.

It is next assumed that just enough of the condensate evaporates to saturate the mixture. The heat required for this evaporation is

$$dQ_2 = L(w_s - w')dm,$$

where w' is the mixing ratio of the entrained air.

Condensation occurs during the ascent however, releasing an amount of heat given by

$$dQ_3 = -mLdw_s.$$

During the process, therefore, the cloudy parcel loses heat in the amount $dQ_1 + dQ_2$ and gains heat in the amount dQ_3. From the first law of thermodynamics,

$$m\left(c_pdT - R'T\frac{dp}{p}\right) = -(dQ_1 + dQ_2) + dQ_3.$$

By dividing both sides by mc_pT and employing (1.26), it follows that in this entrainment process

$$\frac{d\theta}{\theta} = -\frac{L}{c_pT}\,dw_s - \left[B + \frac{L}{c_pT}(w_s - w')\right]\frac{dm}{m}. \qquad (4.19)$$

When there is no entrainment ($dm = 0$), this result reduces to an expression for the change of potential temperature in the pseudoadiabatic process, as expected. Because the bracketed term is always positive in cases of interest, (4.19) implies that temperature falls off at a faster rate when entrainment is taken into account than when it is neglected. This means that buoyancy is impaired by entrainment, as was anticipated.

The rate of change of temperature following a cloudy element is obtained by differentiating (4.19) with respect to time. It can be shown that

$$\frac{d}{dt}(T - T') = (\gamma - \Gamma_s)\frac{dz}{dt} - \left[(T - T') + \frac{L}{c_p}(w_s - w')\right]E, \quad (4.20)$$

where $E = (1/m)(dm/dt)$ is the entrainment rate of the outside air, γ is the environmental lapse rate, and Γ_s is the pseudoadiabatic lapse rate.

As an alternative to lateral mixing, Squires (1958b) suggested that cloud dilution might be explained by the entrainment of dry environmental air from the region just above cloud top. As a result of turbulence or other processes, a parcel of ambient air may be drawn into the cloud, causing the evaporation of some of the cloud droplets. This will chill the parcel, reducing its buoyancy, leading to a downdraft. The cumulative effect of many such penetrative downdrafts from cloud top will be to cool and dry the cloud, especially in its upper region.

The equation governing the temperature T_p of the entrained, unsaturated parcel is

$$\frac{d}{dt}(T_p - T) = -E(T_p - T) - E\frac{L\mu}{c_p} - (\Gamma - \gamma_c)\frac{dz}{dt}. \quad (4.21)$$

Here, as in (4.20), T denotes the cloud temperature and $E = (1/m)(dm/dt)$ is the entrainment rate. Γ is the dry adiabatic lapse rate and γ_c is the ambient lapse rate in the cloud. μ is the mixing ratio of the condensed water. The first term on the right represents the warming of the unsaturated parcel by mixing with the warmer cloudy air. The second term is the cooling of the parcel by the evaporation of cloud water. The third term gives the rate of temperature change between the parcel and its cloudy environment caused by the vertical motion of the parcel, dz/dt.

The trajectory of the air parcel as well as the vertical extent and the time scale of the penetrative downdraft can be calculated by including the equations for the parcel velocity and its vapor mixing ratio:

$$\frac{dU_p}{dt} = \left(\frac{T_p - T}{T}\right)g + g\mu - E(U_p - U) \quad (4.22)$$

$$\frac{dw_p}{dt} = -E(w_p - w - \mu). \quad (4.23)$$

In these equations, $U_p = dz/dt$ is the vertical velocity of the parcel, U is the vertical velocity of the cloudy environment through which it descends, w is the vapor mixing ratio of the cloud, and w_p that of the parcel. In (4.22) the first term on the right-hand side is the thermal buoyancy term, the second is the buoyancy created by the water burden of the environment, and the third is frictional deceleration caused by eddy momentum exchange between the parcel and the cloudy environment.

(d) Aerodynamic resistance: the bubble theory, jets and plumes

Pictured in Fig. 4.3 is an idealized thermal, based primarily on laboratory studies of convection, although having a resemblance to atmospheric thermals which appear as protuberances or "turrets" on cumulus clouds.

Such thermals are observed to be shape-preserving. That is, they have a form which maintains geometrical similarity during much of their development. A theory based on dimensional analysis has been formulated to explain some of their gross features. It is argued that the vertical velocity of a bubble depends on its size and the buoyancy according to the particular combination

$$u = c(g\overline{B}r)^{1/2}, \tag{4.24}$$

where u is the upward velocity of the cap, $\overline{B}$ is the average value of the buoyancy factor across the bubble, r is the radius of the cap, and c is a dimensionless constant to be determined experimentally or theoretically. Because of similarity the height of the cap above ground may be expressed as $z = nr$ and the volume of the thermal as $V = mr^3$, with n and m dimensionless constants to be determined. It is assumed that the total

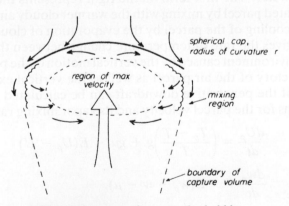

FIG. 4.3. Structure of a convective bubble.

buoyancy (product of V times B) is conserved. Therefore, at any time,

$$V\bar{B} = V_0\bar{B}_0,$$

where V_0 and $\bar{B}_0$ denote the initial volume and buoyancy.

With these assumptions, (4.24) may be written

$$u = cn(gr_0^3\bar{B}_0)^{1/2}/z.$$

Upon integration this result reduces to

$$z^2 = 2cnt\sqrt{\beta/m}, \tag{4.25}$$

where $\beta = gV_0\bar{B}_0$. Laboratory results confirm (4.25), with $m \approx 3, n \approx 4$, and $c \approx 1.2$.

In the atmosphere, cumulus clouds are more complicated than the elementary bubbles, producible in the laboratory, which this theory is designed to explain. The individual spherical turrets in cumulus clouds are strongly suggestive of bubbles, and to a limited extent their behavior is consistent with the elementary theory. Their velocity, however, is intimately connected with the stability of the air and the size and state of development of the cloud as a whole, and cannot be predicted for all clouds and for all occasions with the set of dimensionless parameters given above.

Even the elementary theory gives some insight on the interaction among thermals expected in cumulus clouds. Owing to the $\sqrt{t}$ law on height and size, successive thermals from the same place will tend to merge with one another, effectively increasing the buoyancy and rate of ascent of the (composite) thermal. If thermals do follow closely one after the other, the convection then takes the form of a continuous jet or plume instead of discrete bubbles. Again, laboratory experiments supported by a large body of theoretical work have been devoted to plumes, which are thought to approximate in some respects the airflow in developing cumulonimbus clouds.

An idealized plume is shown in Fig. 4.4. Its shape is conical, and the profiles of velocity and buoyancy across the plume are indicated. Since buoyancy and temperature are related through the factor B, the buoyancy profile is essentially the same as a temperature profile. Because of the conical shape, the radius is expressible as

$$r = az. \tag{4.26}$$

The mass flux through a given level is equal to $AUr^2\varrho$, with units of kg/s, with A a dimensionless factor determined by the shape of the velocity profile. The momentum flux is $AU^2r^2\varrho$, with units of kg m s^{-2}, the same as force.

The flux of buoyancy is $c\varrho gBUr^2$, with units of kg m s^{-3}, the same as force per unit time. c is a shape factor determined by the profiles of

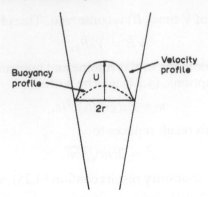

FIG. 4.4. Vertical cross-section of plume.

velocity and buoyancy. Of some theoretical interest is the pure buoyant plume, defined by

$$c\varrho g B U r^2 = \text{const.}, \tag{4.27}$$

in which the buoyancy flux is constant over height. (The shape factor c is assumed constant.)

The buoyancy force and momentum are related in the following way: within a unit height interval the net buoyancy is $c\varrho g B r^2$ and the momentum is $A U r^2 \varrho$. The buoyancy in this slice of air is equal to the time rate of change of momentum. Thus

$$c\varrho g B r^2 = \frac{d}{dt} (A U r^2 \varrho) = U \frac{d}{dz} (A U r^2 \varrho). \tag{4.28}$$

Together, (4.27) and (4.28) may be employed to deduce the dependence upon height of the velocity and the buoyancy in the plume. It is assumed that these quantities are proportional to some power of height: $U \propto z^a$, $B \propto z^b$. Then the only way of satisfying (4.27) and (4.28) is for $a = -1/3$ and $b = -5/3$. Thus, for a pure buoyant plume, theory predicts

$$U \propto z^{-1/3}; \qquad B \propto z^{-5/3}. \tag{4.29}$$

The elementary theory presented here is for a dry plume, in which condensation is not considered. The closest atmospheric analogue to a plume is the updraft in cumulonimbus clouds, where condensation is occurring. The theory has been elaborated by several investigators to include the effects of condensation, and with these refinements it comes closer to describing natural convection. However, it still contains a number of adjustable dimensionless parameters that are not found to be the same under all conditions.

The most recent work on cloudy convection has not focused attention on the updraft only, nor employed arguments based on dimensional analysis, but has attempted to integrate the equations of motion of air while including the thermodynamics of condensation and precipitation production. Such dynamic models show the development of cloud and rain that closely resembles natural convection. Some of the recent models of cloudy convection are described in Chapter 15.

Problems

4.1. Show that the change of adiabatic liquid water mixing ratio with height can be written

$$\frac{d\chi}{dz} = \frac{c_p}{L} (\Gamma - \Gamma_s).$$

4.2. Show that the height H of the lifting condensation level is related approximately to the temperature and dew point of the air at the surface by

$$H \text{ (in km)} \approx (T - T_d)/8.$$

4.3. Derive an expression for the change of relative humidity with respect to height. Prove that in a well-mixed, unsaturated layer of air the relative humidity will always increase with height. Assuming a well-mixed layer, evaluate the expression for $T = 250°K$ and $f = 50\%$.

4.4. In cloudy convection the mass of condensed water that is carried along with an ascending parcel of air exerts a downward force on it and thus reduces its buoyancy. This effect may be described by applying a correction factor to the parcel's temperature to account for the effect of condensed water on the buoyancy. (This is analogous to the virtual temperature correction to account for water vapor.) Derive an expression for the correction factor in terms of the mass of condensate present.

4.5. Equal masses of two samples of air are thoroughly mixed and a fog is observed to form. The first sample has a temperature of 30°C with 90% relative humidity; the second a temperature of 2°C with 80% relative humidity. Assuming that the mixing occurs isobarically at 1000 mb, determine the temperature of the foggy air and its liquid water content, in grams of water per kilogram of air. Saturation vapor pressure as a function of temperature is given in the following table:

T(°C)	e_s(mb)
0	6.11
2	7.06
4	8.14
6	9.35
8	10.73
10	12.28
12	14.03
14	15.99
16	18.18
18	20.64
20	23.39
22	26.44
24	29.85
26	33.63
28	37.82
30	42.45

4.6. The following sounding was obtained in central Alberta at a time when thunder-storms were developing.

Pressure (mb)	Temperature (°C)	Dew point (°C)
910 (surface)	23.5	14.5
850	17.0	12.5
800	12.5	10.8
770	10.0	6.0
745	10.0	−1.5
660	2.0	−10.0
555	−10.0	−13.0
525	−12.0	−19.0
500	−13.0	−18.0
400	−24.5	−30.5
300	−39.5	—
215	−55.0	—
190	−53.0	—
180	−49.0	—
125	−53.0	—

(a) Consider the air at surface level. For this air, determine the following quantities:

(1) potential temperature
(2) density
(3) mixing ratio
(4) relative humidity
(5) virtual temperature
(6) equivalent temperature
(7) equivalent potential temperature
(8) wet-bulb temperature
(9) wet-bulb potential temperature
(10) isentropic condensation temperature.

(b) Suppose a parcel of air from 910 mb ascends adiabatically.

(1) At what pressure would condensation be reached?
(2) Assuming ascent to continue beyond the condensation level, what would be the adiabatic liquid water content at 500 mb?
(3) What would be the increase in entropy of the parcel by the time 500 mb is reached?
(4) What would be the amount of latent heat released by the time 500 mb is reached?
(5) Estimate the vertical velocity of the parcel at 500 mb, neglecting friction and the mass of condensed water.

(c) What is the amount of columnar water vapor between the surface and 500 mb?

4.7. For a pure buoyant plume, it can be shown that the vertical velocity and the buoyancy factor of the ascending air vary with time according to $U \propto t^{\alpha}$ and $B \propto t^{\beta}$. Solve for the numerical values of α and β.

4.8. The specific enthalpy of air is defined in Problem 1.5 by $h = u + p\alpha$.

(a) Show that h is conserved in an adiabatic, isobaric process.
(b) Assume that a saturated cloudy parcel, consisting of m_d kilograms of dry air, m_v kilograms of water vapor, and m_l kilograms of liquid water, constitutes a closed thermodynamic system. Show that its specific enthalpy is

$$h = h_d + w_s h_v + \chi h_w$$
$$= (c_p + c_w Q)T + w_s L,$$

where h_d, h_v, and h_w are respectively the specific enthalpies of dry air, water vapor, and liquid water, and $Q = w_s + \chi$.

(c) By making use of the results in (a) and (b) and the inequalities $c_w Q / c_p \ll 1$ and $w_s L / [T(c_p + c_w Q)] \ll 1$, show that the wet equivalent potential temperature (2.43) can be approximated by the following relation in an isobaric, adiabatic process:

$$\theta_q = Ch(1 - aQ)(1 - bQ)$$
$$\approx Ch[1 - (a + b)Q],$$

where

$$C = \frac{1}{c_p} \left(\frac{100 \text{ kPa}}{p_d} \right)^k, \quad a = kb \ln \left(\frac{100 \text{ kPa}}{p_d} \right), \quad \text{and} \quad b = \frac{c_w}{c_p}, \quad \text{with} \quad k = R'/c_p.$$

(d) Assume that x kilograms of air, having wet equivalent potential temperature θ_{q1} and total mixing ratio Q_1, mixes isobarically with y kilograms of air, with corresponding properties θ_{q2} and Q_2. For $x + y = 1$, show that the wet equivalent potential temperature of the mixture is given approximately by

$$\theta_q \approx x\theta_{q1} + y\theta_{q2},$$

subject to the requirements

$$|Q_1 - Q_2| \le 10 \text{ g/kg}$$

and

$$|\theta_{q1} - \theta_{q2}| \le 15 \text{ K},$$

which are usually satisfied in the atmosphere.

4.9. An unsaturated parcel of air is entrained into a cloud near the region of the cloud top. The updraft velocity U, the ambient lapse rate in the cloud γ_c, the entrainment rate E, and the mixing ratio of the condensed water μ can be considered constants along the trajectory of the parcel. Assuming that at $t = 0$ the velocity of the parcel is U_p and its acceleration is zero, obtain from (4.21) and (4.22) analytical solutions for the evolution in time of the parcel's velocity, temperature, and height. For simplicity, assume that the term $g(T_p - T)/T$ in (4.22) can be approximated by $g(T_p - T)/T_0$ where T_0 is a constant average temperature in the cloud. Solve also for the time it takes for the parcel velocity to reach its maximum value. Show that for $T_0 = -15°C$ and $\gamma_c = 6°C/\text{km}$, this time equals approximately 4.3 minutes.

5

Observed Properties of Clouds

THE visual appearance of the planet Earth when viewed from space is dominated by clouds and cloud patterns. Clouds exist because of the physical process of condensation, but condensation occurs mainly in response to dynamical processes that include widespread vertical air motions, convection, and mixing. Accordingly, the pattern and structure of clouds are influenced by dynamical factors such as stability, convergence, and the proximity of fronts and cyclones. But this is not to say that clouds are merely incidental to atmospheric motions; for indeed clouds affect the motions through physical processes including the release of latent heat, the redistribution of atmospheric water and water vapor, and the modulation of the transfer of solar and infrared radiation in the atmosphere. In short, cloud behavior is dominated by dynamics but any comprehensive view of atmospheric dynamics must include clouds.

Central to the subject of cloud physics are the topics of drop formation, growth, and interaction. These are treated in later chapters. As a background to the theory of cloud and precipitation development, the present chapter describes some of the observed characteristics of clouds.

Sizes of clouds and cloud systems

Satellite views of the Earth, such as Fig. 5.1, reveal organized cloud patterns extending over distances of hundreds and thousands of kilometers. Some of these patterns are associated with low-pressure centers at the surface and their accompanying fronts; others are tied to orographic features. They move with the pressure pattern or remain in place, and can continue to exist for days at a time.

The individual water droplets and ice crystals that constitute a cloud are transitory, created by condensation and lost by evaporation or precipitation. The cloud or cloud system continues to live by the steady creation of new droplets as the previous ones cease to exist. It may be thought of as a kind of wave or disturbance through which the droplets move. It propagates in the direction of new drop formation and does not necessarily move as an entity with the wind at cloud level. The individual

FIG. 5.1. View of Earth from the first Geostationary Operational Environmental Satellite (GOES-1). (Photo courtesy of NASA.)

droplets, crystals, and precipitation particles, however, move with the
velocity of the air surrounding them plus their fall velocity relative to the
air.

Within a cloud system there can be recognizable isolated clouds or
cloud elements that are identifiable by their shape and size. These might
range from 1 to 100 km in linear extent and have lifetimes from minutes
to hours. Together, they make up the cloud system.

Closer views, as in Fig. 5.2, show increasing detail in cloud structure,
down to the minimum resolvable scale. To assist in organizing thought
and language, it has been found convenient to introduce a hierarchy of
scales to describe atmospheric phenomena. Phenomena of the largest
scale, variously called the masoscale, macroscale, or synoptic scale,
range from about 1000 km upwards, and include the cloud systems

FIG. 5.2 (a). Streams of convective clouds ("cloud streets") forming between Labrador and Greenland as cold air flows from the north over the warmer open water. A mesoscale cyclone is developing in the center of the photograph. (Courtesy of John Lewis and Atmospheric Environment Service of Canada.)

associated with extratropical cyclones (low-pressure centers) and fronts. At the other extreme, phenomena of scale about 1 km or less are called microscale. These include small and generally fleeting cloud "puffs", the turrets discernable on cumulus clouds, and small scale cloud irregularities, as well as physical processes on the scale of a cloud droplet (microphysical processes). In between, with scales from a few kilometers to several hundred kilometers, are the so-called mesoscale phenomena, which include many kinds of clouds and cloud systems.

Clouds are classified according to their visual appearance from the ground in a system proposed by Luke Howard in 1803 and now adopted internationally. Howard, a British pharmacist, optimistically predicted that by his classification meteorology would be "rescued from empirical mysteriousness, and the reproach of perpetual uncertainty". Unfortunately, meteorology has remained a subject with more than its share of

FIG. 5.2 (b). Mixture of cloud types photographed over the tropical ocean from Gemini 4 spacecraft. (Courtesy of NASA.)

difficult problems, though Luke Howard's classification has served dutifully to this day. The system distinguishes four major cloud types: *cumulus* (clouds with vertical development), *stratus* (clouds in flat-appearing layers), *cirrus* (fibrous or hair-like), and *nimbus* (rain clouds). The complete international classification includes dozens of cloud types that may be some combination of the four major divisions (e.g., cirro-cumulus, nimbostratus), types distinguished by their altitude (alto-stratus), or others notable for their development (cumulus congestus). It is assumed that the reader is acquainted with the main categories of cloud classification. Many cloud atlases are available with excellent illustrations, for example Scorer and Wexler's (1963) *A Colour Guide to Clouds*.

Stratus clouds can extend over hundreds of kilometers in the horizontal, and can be thin and non-precipitating or thick enough to produce substantial widespread rain or snow. They are formed by synoptic or mesoscale vertical air motion arising from large scale convergence, frontal lifting, or orographic lifting of air that is statically stable. Stratus cloud at the earth's surface is called fog, and may be caused by radiational cooling of the air near the ground, or by the mixing of air masses having widely different temperatures, as in coastal regions.

Cumulus clouds are caused by convection in unstable air. Their horizontal and vertical extents are comparable, and the vertical extent is controlled by the depth of the unstable layer and its degree of instability. A typical length scale of a cumulus cloud is 3 km, though these clouds start from smaller thermals in which condensation first occurs, and can grow to extend vertically throughout the troposphere, actually penetrating a few kilometers into the stratosphere. Clouds this large invariably produce rain and usually lightning and thunder, and are called cumulonimbus clouds. Cumulonimbus clouds that last longer than an hour or so continue to spread horizontally and can eventually reach a size of 100 km or more. They are of special interest because of the severe weather that often accompanies them: hail, flood, and high winds. Cumulus clouds that produce rain without reaching thunderstorm proportions are called showers.

Microstructure of cumulus clouds

(a) General aspects and experimental procedures

Parcel theory suggests that the microstructure of cumulus clouds is mainly a function of height. However, early observations of convective cloud structure in the 1940s indicated that there is considerable spatial and temporal variability in cloud properties. Vertical motions and air temperature recorded by aircraft in the Thunderstorm Project (Byers and Braham, 1949) revealed that a convective storm usually consists of a number of "cells", each of which passes through a characteristic life cycle. At any moment, a storm ordinarily contains cells at different stages of development. But also within a cell there are smaller scale fluctuations. Measurements of updraft speed, temperature, and liquid water content typically show a detailed fine structure with significant variations occurring over a distance of a few meters. Often, the fluctuations of updrafts, temperature, and water content are well correlated, indicating the importance of feedback mechanisms between dynamic, thermodynamic, and microphysical processes in convective clouds. In spite of the large natural variability, the "average" microstructure of convective clouds is largely controlled by a few characteristics of the environmental air. These factors are usually the cloud base temperature, the type and concentration of condensation and ice nuclei, the stratification of temperature and humidity, and the amount of dynamic forcing by vertical windshear and large scale convergence.

The microstructure of cumulus clouds is usually described in terms of temperature, vapor content, vertical velocity, liquid water content, cloud droplet spectra, and the size distributions of raindrops and ice particles. During the last 40 years much effort has been devoted to develop instruments and experimental techniques to measure these quantities. Progress has been made on radar, lidar, and other remote sensing techniques for measuring cloud properties. However, the most reliable technique (although expensive) is still to take measurements in-situ using research aircraft equipped with the necessary sensors and instrumentation. The sampling procedure usually consists of a series of horizontal traverses flown through the cloud core at different altitudes.

A detailed description of the instrumentation used on research aircraft, though an important part of cloud physics, is not included here. For this information the reader must turn to reports dealing with field experiments on convective clouds. We will only mention some of the most common instruments for measuring microphysical parameters. Temperature is ordinarily measured by fast response thermometers such as the Rosemount sensor. Humidity can be recorded with a Cambridge dewpoint hygrometer. Cloud liquid water concentration is measured by a Johnson–Williams meter, which deduces the water amount from the heat lost by an exposed hot wire due to the evaporation of the droplets it intercepts. Vertical velocities are computed from measurements recorded by an accelerometer, a rate gyro system, and sensors for air speed, attack angle and sideslip angle. Cloud droplet spectra can be measured with an instrument called a forward scattering spectrometer probe built by PMS, Inc. This instrument sizes and counts particles by sensing the amount of light scattered as the particles interact with a focused laser beam. Size distributions of raindrops are measured by a similar type of optical array spectrometer. Information about the shape and thermodynamic phase of particles is gained by using the PMS two-dimensional imaging probe. This instrument records shadow images cast by particles passing through a laser beam. A high speed memory enables the instrument to collect and synthesize information about every 350 ns.

(b) Variation of cloud properties with height

The typical variations of cloud microstructure with height will be discussed with reference to a set of observations in a particular cloud. The data were recorded by scientists of the Atmospheric Environment Service of Canada (Schemenauer *et al.*, 1980) during their participation in the High Plains Experiment conducted in Miles City, Montana. The

FIG. 5.3. Photograph of cumulus cloud taken at 2250 GMT on July 19, 1979, near Miles City, Montana. This is the cloud that was sampled for some of the illustrations that follow. (Photo courtesy of George Isaac and AES.)

cloud we will describe was observed at 2300 GMT on July 19, 1979. As can be seen from the photograph of Fig. 5.3, it had a typical cumulus appearance with a clear-cut, cauliflower-shaped cloud top. Fairly isolated from other clouds, it was easy to identify over the 10-min period required for sampling. The cloud base was at 3.8 km, 1.2°C, and 63.55 kPa. The cloud thickness was 1.5 km and remained approximately constant during the sampling period. Cloud parameters were measured during five horizontal traverses starting at cloud top and in steps 300 m apart in altitude. The data from the two-dimensional imaging probe indicated that no ice particles or raindrops were present.

Figure 5.4 displays the measurements of liquid water content and vertical velocity for the penetration flown at an altitude of 4.58 km. Because the flight speed was approximately constant at 70 m/s, this record corresponds to a distance of about 2.2 km. The data indicate variability on a small scale. High liquid water contents are clearly associated with strong updrafts. This correlation, though obvious in this traverse, is not always observed. An interesting feature of the vertical velocity measurements is that downdrafts exist over a significant fraction of the traverse. They are particularly pronounced near the cloud edges. The presence of in-cloud downdrafts, a finding that is typical for cumulus clouds, cannot readily be explained by simple parcel theory.

For each of the five cloud penetrations, the maximum values of updraft

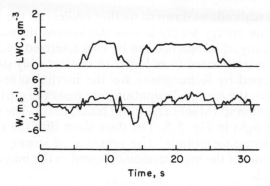

FIG. 5.4. Measurements of water content and updraft velocity in the cloud of Fig. 5.3 along a track at altitude of 4.58 km. (Adapted from Schemenauer *et al.*, 1980.)

and downdraft (measured over time intervals of 0.5 s), together with the root mean square (RMS) vertical velocities, have been computed. Figure 5.5 shows the results plotted as functions of height above cloud base. The peak gusts are found to be stronger in the upper third of cloud. Maximum values of updraft and downdraft are comparable in magnitude. The values of the RMS vertical velocity are about half those of the peak drafts. The variability of the vertical air velocity indicates the importance of turbulence in the cloud. Turbulence is usually pictured as the flow of

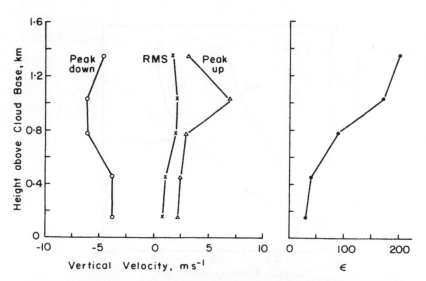

FIG. 5.5. Maximum observed updrafts and downdrafts and RMS vertical velocity plotted against height above cloud base. Also shown is the computed turbulent energy dissipation rate, ε, in units of $cm^2 \, s^{-3}$.

energy from large scale eddies down to smaller eddies. At the end of this energy cascade, the energy finally leaves the system through viscous heating. The intensity of turbulence can be characterized by the rate at which the energy is transferred from larger to smaller eddy sizes. Using the theory developed by Kolmogorov for the inertial subrange, it is possible to estimate the energy dissipation rate from the gust measurements sampled by aircraft (MacPherson and Isaac, 1977). The results are plotted at the far right in Fig. 5.5. The data show that the turbulence intensifies with increasing altitude. The presence of strong turbulence can account for some of the high variability found in the microstructure of convective clouds.

Turning now to the liquid water content structure, we have plotted penetration averages and maximum 0.5-s values as functions of height above cloud base in Fig. 5.6. The calculated adiabatic liquid water content is also shown. The average liquid water content increases with height, but is only about half of the full adiabatic value. The maximum 0.5-s liquid water contents (that is, the maxima obtained from averages over about 35 m) are close to adiabatic, which might indicate the presence of undiluted "adiabatic pockets". However, the full adiabatic values are realized only in regions that are small in relation to the cloud as a whole, and the main body of the cloud has significantly less water than the adiabatic value. This finding is common for cumulus clouds. The only plausible explanation is that mixing must have occurred between the

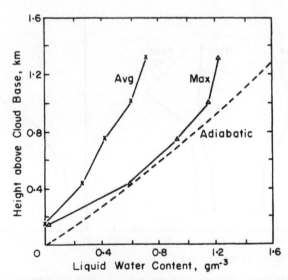

FIG. 5.6. Maximum, average, and adiabatic liquid water contents plotted against height above cloud base. (Adapted from Schemenauer *et al.*, 1980.)

rising air from cloud base and the dry environmental air. This process will be discussed later in some detail. For the cloud described here, there was further evidence of cloud–environment interaction. The observed in-cloud temperature values (not shown) were less than those predicted by pseudoadiabatic ascent, suggesting that some cool environmental air must have entered the cloud.

Cloud droplet distributions were sampled for drops between 2 and 30 μm in diameter with a resolution of 2 μm. Penetration averages of the size spectra are displayed at the corresponding flight levels in Fig. 5.7. Also plotted are the mean droplet concentration and the average droplet diameter as functions of height above cloud base. The most significant observation from Fig. 5.7 is the increase of mean droplet diameter from 7–9 μm near cloud base to 13–14 μm near cloud top, which is consistent with the shift of the droplet spectrum to larger sizes at higher altitudes. In the middle and upper portions of the cloud the droplet concentration decreases with increasing altitude. This suggests that some of the cloud droplets, while being lifted in the updraft, might have collided and joined with each other to form fewer but larger drops. The observations show that the droplet distributions evolve during ascent, but it is impossible from the data available to speculate whether the growth processes are adequate in this particular cloud to lead eventually to the formation of precipitation.

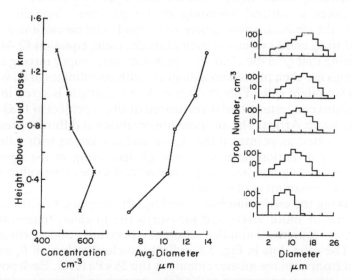

FIG. 5.7. Variation with height of the average droplet concentration and the average droplet size. The form of the average droplet spectrum at each height is also shown. (Adapted from Schemenauer *et al.*, 1980.)

It is important to recognize that the liquid water in a cloud is relatively small in amount and highly dispersed. The mass of water in a cloud is in the order of 0.1% or less of the mass of the air. The volume of a cloud occupied by condensed water is only in the order of one part in 10^6. Because this water is spread over many small droplets, it has a high surface-to-volume ratio, which partly explains the significance of clouds in chemical processes in the atmosphere.

(c) Observational evidence of cloud top mixing

Observations in developing cumulus clouds often reveal temperatures and water contents less than the adiabatic values, consistent with dilution by mixing of some sort. The origin of the entrained air is a matter of much debate. Two different explanations have been offered, which depend on the theory of convection adopted. If clouds are considered to resemble jets or plumes, outside air is entrained through the sides and mixes in the updraft in a way similar to the results of laboratory tank experiments. On the other hand, in the bubble or thermal theory, air mixes in near the top of the cloud. This mixing causes some of the cloud water to evaporate, chilling the air. The cooled mixture can then descend into the cloudy air below in the form of penetrative downdrafts.

The wet equivalent potential temperature θ_q and total water mixing ratio Q can be examined to infer the origin of the entrained air in a non-precipitating cloud. In a procedure developed by Paluch (1979), an airplane takes a vertical sounding of temperature, humidity, and pressure in the clear-air environment of a cloud. The measurements are converted to θ_q and Q values at each altitude, using equation (2.44) for θ_q and noting that Q in the clear air equals the water vapor mixing ratio. These values are then plotted on a diagram with coordinates of θ_q versus Q (which we might call a Paluch diagram). An example is given in Fig. 5.8, a cumulus congestus cloud investigated during a project in 1983 near Nelspruit, South Africa. The airplane observations at different pressure levels are plotted as points on the figure and connected with a line as shown. The convention here is to plot Q decreasing in the upwards direction on the diagram, because Q ordinarily decreases with altitude in the clear-air environment of a cloud.

After taking the environmental sounding, the airplane penetrates the cloud at one or more levels and measures the in-cloud temperature, liquid water content, humidity, droplet spectrum, and vertical air velocity. The solid dots in Fig. 5.8 indicate in-cloud values of θ_q and Q calculated from airplane measurements at the 38 kPa level. Each point is based on a one-second average of the data, corresponding to a horizontal distance of 150 m. The points plotted here were obtained consecutively as the airplane flew a distance of 1.2 km within the cloud.

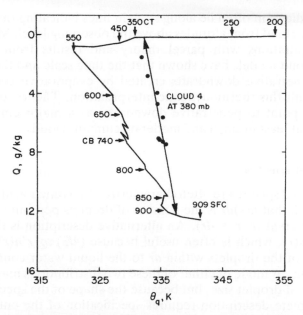

FIG. 5.8. Environmental sounding, plotted on coordinates of Q versus θ_q (a Paluch diagram). Cloud base (CB) and cloud top (CT) are indicated. Consecutive measurements at one level in the cloud are shown by points. (From Reuter and Yau, 1987a.)

Because thermodynamic processes in a non-precipitating cloud may be regarded as reversible, saturated, and approximately adiabatic, θ_q and Q are conservative parameters that mix linearly according to (4.7) and (4.8). Therefore, any cloudy sample consisting of air from two discrete levels will have a point on the Paluch diagram that lies on the straight line connecting the characteristic points of these levels. Obviously, a single in-cloud point can be the result of mixing from any source regions. However, if the points obtained from consecutive samples at a given observational level lie along a straight line, it is highly probable that the region sampled consists of a mixture of the air from two discrete sources.

The cloud points in Fig. 5.8 lie approximately on a segment of the straight line connecting environmental points from near the ground and from close to cloud top. This is strong evidence for penetrative downdrafts as the main mixing mechanism at the 38 kPa level. If the dilution of the cloud had been caused by air entrained from a lower level, the points would have been aligned differently.

A body of observational evidence is accumulating, which supports the idea that cloud-top mixing and penetrative downdrafts account for much of the dilution of small and medium-size cumulus clouds. In-cloud points on a Paluch diagram often lie along straight lines indicating that the dry air enters the cloud from altitudes above the observing level. Moreover, simple computations with parcel theory and results from complex numerical cloud models have shown that the time scale and the vertical extent of penetrative downdrafts created by evaporative cooling are consistent with this thermodynamic interpretation. Theory and observations thus point to penetrative downdrafts as a major entrainment mechanism, at least in small and moderate cumulus clouds.

Cloud droplet spectra

Cloud droplet spectra are often characterized by a function $n(r)$, which is defined such that $n(r)dr$ is the number of droplets per unit volume in the radius interval $(r, r + dr)$. An alternative description is the distribution of $r^3n(r)$, which is often useful because $(4/3)\pi\varrho_L r^3 n(r)dr$ is the contribution of the droplets within dr to the liquid water content. The mean, median, or mode of either of these two distributions may be used as a measure of droplet size, but because the shape of the spectrum can vary, a complete description requires specification of the entire spectrum.

Average droplet spectra obtained by combining a large number of samples tend to exhibit a characteristic shape: $n(r)$ rises sharply from a low value to a maximum and then decreases with increasing droplet size, creating a distribution that is positively skewed with a long tail toward the larger sizes. Such a distribution can be approximated by either a lognormal or a gamma distribution function. Of course, individual size spectra may differ substantially from this shape. Many spectra have been observed that show only weak positive skewness and some are even negatively skewed (Squires, 1958a). Also, many individual spectra are bimodal. Warner (1969a) suggested that the tendency of size distributions to be bimodal is the result of mixing at the cloud top between the cloud and the dry environment.

Squires focused attention on observations that showed that droplet spectra of cumulus clouds in continental air differ greatly from those found in clouds formed over the ocean (Fig. 5.9). Continental cumulus clouds typically have a large concentration of small droplets and a narrow size spectrum, while maritime cumulus clouds have a relative small concentration of large droplets and a broad size spectrum. There are many observational studies (for example Ryan *et al.*, 1972) that confirm that maritime clouds usually have a broader size spectrum than continental clouds. It is significant, however, that the liquid water contents of maritime and continental cumulus clouds are often similar. The marked differences in droplet size spectra are largely accounted for by the different air mass types with correspondingly different concentrations of

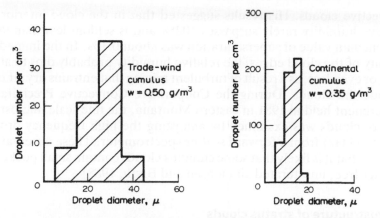

FIG. 5.9. Droplet spectra in trade-wind cumulus off the coast of Hawaii and continental cumulus over Blue Mts. near Sydney, Australia. (From Fletcher, 1962, after Squires, 1958a.)

cloud condensation nuclei, a distinction that will be explained in some detail in later chapters. Suffice it to say for the moment that embryonic droplets are formed by condensation on hygroscopic aerosol particles, and that continental air usually has a higher concentration of such particles than maritime air. Many more droplets compete for the available moisture in continental clouds than in maritime clouds, and consequently they are smaller.

There is a tendency for cloud droplet size spectra to change with time as the drops move about, interacting with one another and with their environment. Usually the spectra broaden as the cloud matures. The broadening is explained by a combination of processes including collisions and coalescence, condensation and evaporation, turbulence effects and the mixing of cloud parcels with different histories. These processes will be described later. At this stage, we only want to mention some of the observed characteristics of the water vapor content of cumulus clouds, because this quantity is crucial for the evolution of droplet distributions.

It is extremely difficult to make accurate measurements of humidity in convective clouds, because of instrument limitations. Humidity sensors usually have a long response time and thus are incapable of measuring the rapid fluctuations in clouds. For example, the Cambridge dew point meter has a response time of about 3 s and cannot adequately record the humidity field inside a turbulent cloud. Information about humidity fluctuations in clouds must therefore be obtained by some other means. Warner (1968) used simultaneous measurements of updrafts and drop-size spectra to estimate the relative humidity in small and moderate-sized convective clouds. His results suggested that in the cloud interior the relative humidity rarely surpasses 102% and is seldom less than 98%. The median value of supersaturation was about 0.1%. In the immediate vicinity of the cloud edges, the relative humidity probably dips to about 90% or even less, as a result of turbulent mixing that entrains dry air from outside the clouds. During the Cooperative Convective Precipitation Experiment held in 1981 in eastern Montana, the fine scale microstructure of clouds was examined by analyzing the high frequency droplet count (50 Hz) from a forward scatter spectrometer. These observations suggest that it is likely that some cumulus clouds contain small pockets of droplet-free, unsaturated air (Jensen and Blyth, 1988).

Microstructure of stratus clouds

In the past, experimental cloud physics has focused more attention on convective clouds than on stratiform clouds. One of the reasons is probably the difficulty of taking representative measurements in widespread cloud layers, which often cover areas of 10^6 km^2 and last for several days. Recently, however, stratiform clouds have come under increasing attention, partly because of their importance in determining the global radiation balance.

The vertical air motions in stratiform clouds are much weaker than those in cumulus clouds. In cumulus, the updrafts and downdrafts are in the order of meters per second; in stratus, they are in the order of a few tens of centimeters per second, values that are consistent with the vertical motion predicted from the horizontal convergence of low-level synoptic scale flow. Though weak, the ascent can continue for a long enough time to produce widespread, continuous rain. Although stratus clouds may appear to be horizontally homogeneous, there is always some degree of fine-scale variability in cloud structure. An example of such variability is shown in Fig. 5.10. These records were obtained by an instrumented airplane as it descended gradually through a marine stratus cloud deck about 200 m thick and many tens of kilometers in horizontal extent. The figure also illustrates typical values of stratus cloud properties.

Liquid water contents of stratus clouds are usually in the range from 0.05 to 0.25 g/m^3. In stratocumulus and deep nimbostratus the values can reach nearly 1 g/m^3. Detailed observations of the microphysical structure of maritime stratus clouds (e.g., Nicholls, 1984; Noonkester, 1984) show that the average liquid water content over horizontal layers increases with height, but the values are less than adiabatic. This increase of water content with height is accounted for by an increase in droplet sizes rather than concentration. In fact, the concentration is observed to be approximately constant throughout much of the cloud, whereas the mean

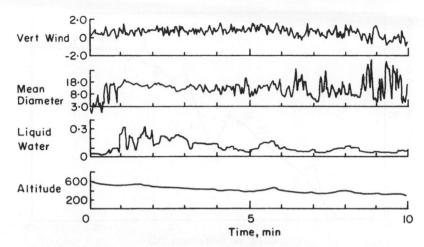

FIG. 5.10. Records of vertical air velocity (m/s), mean droplet diameter (μm), liquid water content (g/m^3), and altitude (m), during descent through marine stratus cloud. (From Telford and Wagner, 1981.)

droplet size increases monotonically with height. This finding points to a growth process dominated by condensation rather than coalescence. Consistent with this interpretation, the droplet spectra in stratus clouds are relatively narrow, as expected from growth by condensation.

Stratus clouds are often capped by a temperature inversion, inhibiting further vertical growth. Turbulent velocity fluctuations near the cloud top cause parcels of dry air to be mixed into the cloud layer. This mixing leads to complete or partial evaporation of some of the droplets near cloud top. The exact nature of this mixing and its effects on the evolution of the droplet size distribution are still debated issues, mentioned earlier and discussed in more detail in Chapter 8. The evaporation causes cooling near cloud top, creating local pockets of air with negative buoyancy that move downwards into the body of the cloud. Significant cooling is also caused by longwave radiative heat loss from cloud top. Nicholls and Turton (1986) evaluated the relative importance of radiative and evaporative cooling at the top of stratiform cloud sheets. They found that evaporative cooling is more important when the turbulent mixing is strong, but that radiation predominates when mixing is weak.

Likelihood of ice and precipitation in clouds

When a cloud of water droplets is cooled to temperatures below 0°C there is a chance that ice crystals will begin to appear. But because the water is highly dispersed and the atmosphere has a relatively short supply of particles that can serve as centers for ice formation (ice nuclei), ice

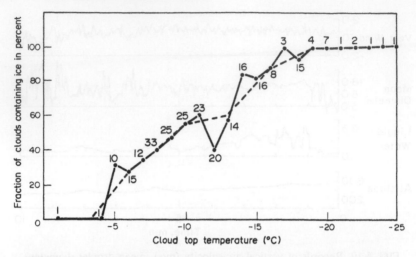

FIG. 5.11. The fraction of clouds containing ice as a function of cloud top temperature, from observations of 258 clouds by several investigators in different regions. The number above each point is the number of observations at that temperature; the dashed curve is a three-point running mean. (From Houghton, 1985, after Hobbs *et al.*, 1974.)

crystals may not be observed until the cloud is cooled to −10°C or colder. That is, cloud water in the atmosphere in the supercooled state is common. Because of the variability in space and time of the concentration of atmospheric ice nuclei, it is not possible to say, for any given cloud, at what temperature the ice phase will appear or how the number of crystals will increase with additional supercooling. Figure 5.11 summarizes results of several cloud studies, indicating the percentage of clouds containing ice as a function of cloud top temperature. Such data may be accepted as an approximate guide to the likelihood of ice, but it should be understood that ice formation depends not only on temperature, but on cloud type, cloud age, and geographical location. Information is not yet available to permit a confident prediction of ice occurrence in a specific cloud.

   Precipitation takes time to develop in a cloud, and is more likely to develop if the cloud is thick and if it contains ice. Therefore the probability of precipitation increases with some combination of cloud age, temperature, and vertical extent. In cumulus clouds, these three quantities are correlated with one another, and the likelihood of precipitation is often related to only one, the cloud thickness. Battan and Braham (1956) reported on the occurrence of precipitation in cumulus clouds at three different locations, the ocean near Puerto Rico, the central United States, and New Mexico. Their data show that the percentage of clouds with precipitation increases with cloud thickness.

The thicknesses for a 20% likelihood are approximately 2.0 km, 3.5 km, and 4.5 km, respectively, for Puerto Rico, the central U.S., and New Mexico. More recent studies have confirmed that continental clouds must be thicker than maritime clouds for the same probability of precipitation. This difference is related to the different cloud droplet populations for the two classes of cloud. In maritime clouds, the droplets are relatively few but large and can collide and coalesce with each other more readily than the small drops in continental clouds to form precipitation. Less is known of precipitation likelihood in stratus clouds, but again the thickness, age, and temperature are the most important controlling factors.

Microstructure of large continental storm clouds

So far our description of cloud microstructure has been confined mainly to small and medium-size clouds. In-situ measurements of cloud parameters are rather scarce for severe convective storms, because often safety requirements do not allow for aircraft traverses through severe storms, particularly those that may contain hail. In addition, large storms occur far less frequently than small cumulus clouds. (Severe storms seem to have the strange habit of "avoiding" the dense observational networks specifically put in place for collecting storm data.) From the observations available, it is clear that the microstructure may vary significantly from one storm to another. Also, within a given storm, there is substantial temporal and spatial variability. Figure 5.12 shows an example of the large spatial variability of vertical velocity in a convective storm, as determined by a series of "dropsondes" released from an overflying airplane. Doppler radar data, presented in a later chapter, also indicate a high degree of variability of the air motion in storms.

In spite of the large variability in storm microphysical parameters, it seems more instructive to discuss measurements in a particular storm rather than to present averages of the properties of many storms. This "case study" approach has the merit of emphasizing the relationships between the different cloud parameters. The data chosen for discussion were obtained in a large multicellular storm that formed near Grover, Colorado, on July 22, 1976, and was studied carefully during the National Hail Research Experiment (Knight and Squires, 1982). The storm is considered fairly representative of large storms forming in an unstable continental airmass. During its 90-min lifetime it produced copious amounts of precipitation including some hail.

Heymsfield and Musil (1982) presented measurements taken by a specially armored airplane at midlevels. Figure 5.13 shows the observations from one of the penetrations. The strongest observed updraft, 26 m/s, was about 2.5 times larger than the peak downdraft speed.

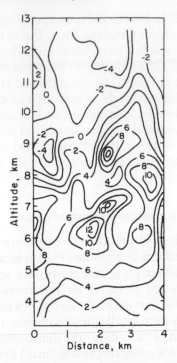

FIG. 5.12. Vertical cross section of vertical air velocity (m/s) in a thunderstorm obtained from a series of dropsonde measurements. (From Bushnell, 1973.)

Considerable variability of the equivalent potential temperature, θ_e, was observed throughout the penetration. It should be noted that the values of θ_e computed from measurements made by an airplane flying below the cloud base were consistently higher than the values at midstorm level, indicating entrainment of environmental air throughout the updraft region. The temperature excess within the updraft regions can be estimated by comparing measured temperatures in the updraft with those outside the cloud at the same level. The peak temperature excess for the data shown is 2°C. The highest values of liquid water content were found in the region of strongest updraft. The peak value of 1.95 g/m³ is about 60% of the computed adiabatic value, 3.2 g/m³. Droplet concentrations (not shown) averaged 750–800 cm⁻³ and the mean droplet diameter ranged between 15 and 18 μm. The ice content, computed from measured size spectra and derived particle habits over the entire size spectrum, was typically less than 1 g/m³. The highest ice concentration was in the region of downdraft. The low ice content in the peak updraft region suggests that depletion of liquid water by growing ice particles was negligible within the updraft core. No particles larger than 5 mm in

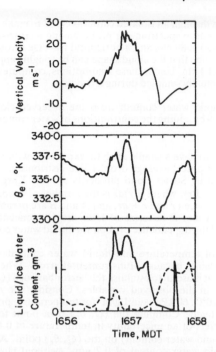

FIG. 5.13. Records of vertical air velocity, equivalent potential temperature, liquid water content (solid), and ice content (dashed) in the core of a
thunderstorm. (From Heymsfield and Musil, 1982.)

diameter were found in the updraft core, whereas comparatively high
concentrations of large particles were found in the downdraft. The
combined liquid and ice amounts in the updraft were lower than the
adiabatic water content by at least 1 g/m³, indicating dilution by entrainment. There is evidence from other studies that some convective storms
are so large in horizontal extent that they contain substantial interior
regions in which the water concentration is undiluted by mixing—so-
called adiabatic cores.

Problems

5.1. Cloudy air consists of dry air, water vapor, and cloud droplets. In a particular cloud
volume, there are 200 droplets per cubic centimeter, all of the same size, with a radius
of 10 μm. The temperature is 10°C and the pressure is 80 kPa. Determine the
following properties of the cloud:

(a) The mass of cloud water per unit volume.
(b) The mass of water vapor per unit volume.
(c) The mass of dry air per unit volume.
(d) The mean distance between the droplets.

5.2. An airplane flies through an isolated cumulus cloud to make microphysical measurements with a raindrop spectrometer probe having a cross-sectional area of 20 cm². The cruising speed is 80 m/s and a horizontal traverse through the cloud core takes about 2 min. Assume that the cloud shape can be roughly approximated by a cylinder having a depth of 4 km. Under these assumptions, compute the fraction of the cloud volume that is actually sampled during a traverse.

5.3. Estimate the liquid water content from the observed cloud droplet size spectra plotted in Fig. 5.9 by adding the contributions of water content from the different size classes.

5.4. It is often useful to have a simple analytic expression with few free parameters that closely fits the empirical average droplet size distribution. A convenient representation, suggested by Soviet cloud physicists A. Kh. Khrgian and I. P. Mazin, is $n(r) = Ar^2 \exp(-Br)$, where $n(r)dr$ is the number of drops per unit volume in the radius interval between r and $r + dr$, and A and B are parameters (Borovikov *et al.*, 1963). Express A and B in terms of total droplet concentration and the average radius. Also find the dependence of A on the liquid water content.

5.5. Measurements of temperature and liquid water content in clouds contain some degree of uncertainty because of instrumental errors. When analyzing data on the Paluch diagram, it is useful to establish the "error bar" of a (Q, θ_q) data point in terms of uncertainties in the observed variables. Consider the measurements of an air sample to be $-10°C$, 60 kPa, and 1 g/m³ for temperature, pressure, and liquid water content, respectively. Compute the (Q, θ_q) data point for these measurements. Compute the effect of an uncertainty in temperature of 0.5°C (without changes in pressure and liquid water content) on the (Q, θ_q) point. Also find the effect of an increase in liquid water content of 0.2 g/m³ (without changing temperature and pressure).

6

Formation of Cloud Droplets

General aspects of cloud and precipitation formation

Phase changes of water are basic to cloud microphysics. The possible changes are as follows:

vapor $\rightleftharpoons$ liquid   (condensation, evaporation)
liquid $\rightleftharpoons$ solid   (freezing, melting)
vapor $\rightleftharpoons$ solid   (deposition, sublimation)

The changes in the direction from left to right in this pattern correspond to increasing molecular order. These transitions do not occur at thermodynamic equilibrium, but in the presence of a strong free energy barrier. Water droplets, for example, are characterized by strong surface tension forces. For a droplet to form by condensation from the vapor, the surface tension must be overcome by a strong gradient of vapor pressure. It is a fact of cloud physics that the phase transitions of most interest, the cloud forming processes, are those that do not occur at equilibrium.

The Clausius–Clapeyron equation describes the equilibrium condition for a thermodynamic system consisting of bulk water and its vapor. Saturation is defined as the equilibrium situation in which the rates of evaporation and condensation are equal. For small droplets, however, because of the free energy barrier, phase transition does not generally occur at the equilibrium saturation of bulk water. In other words, if a sample of moist air is cooled adiabatically to the equilibrium saturation point for bulk water, droplets should not be expected to form. In fact, water droplets do begin to condense in pure water vapor only when the relative humidity reaches several hundred percent!

The classical problem of cloud physics is to explain why cloud droplets are observed to form in the atmosphere when ascending air just reaches equilibrium saturation. The answer has been known for nearly a century, and rests in the fact that the atmosphere contains significant concentrations of particles of micron and submicron size which have an affinity for water and serve as centers for condensation. These particles are called condensation nuclei. The process by which water droplets form on nuclei from the vapor phase is called heterogeneous nucleation. The formation of droplets from the vapor in a pure environment, which requires high supersaturations and is not important in the atmosphere, is called homogeneous nucleation. All processes in which a free energy barrier must be overcome, such as vapor to liquid or liquid to ice transitions are termed nucleation processes.

Many different types of condensation nuclei are present in the atmosphere. Some become wetted at relative humidities less than 100% and account for the haze that impedes visibility. The relatively large condensation nuclei are those that may grow to cloud droplet size. As moist air is cooled in adiabatic ascent, the relative humidity approaches 100%. The more hygroscopic members of the nucleus population then begin to serve as centers of condensation. If ascent continues, supersaturation is produced by the cooling and is depleted by the condensation on nuclei. By supersaturation is meant the excess of relative humidity over the equilibrium value of 100%. Thus air with a relative humidity of 101.5% has a supersaturation of 1.5%. In clouds, there are usually enough nuclei present to keep the supersaturation from rising much above about 1%. It is an important characteristic of the atmosphere that there are always condensation nuclei present to provide for cloud formation when the relative humidity barely exceeds 100%.

As a cloud continues to ascend, its top may eventually be cooled to temperatures below 0°C. The water droplets in the cloud are then said to be supercooled, and they may or may not freeze, depending upon whether ice nuclei are present. For pure water droplets, homogeneous freezing does not occur until a temperature of about −40°C is reached. When suitable nuclei are present, however, freezing can occur at just a few degrees below zero. Although these aerosols are not completely understood, it is significant that they are rather scarce in the atmosphere, unlike the abundant condensation nuclei. Consequently supersaturations of more than a few tenths of a percent are extremely uncommon in the atmosphere, although water droplets in supercooled form are the regular state of affairs. Supercoolings down to −15°C or colder are not uncommon. For this reason one of the main methods of artificially modifying clouds is the addition of ice nuclei, as we shall see later.

A cloud is an assembly of tiny droplets usually numbering several hundred per cubic centimeter and having radii of about 10 μm. This structure is extremely stable as a rule; the droplets show little tendency to come together or to change their sizes except by general growth of the whole population. Precipitation develops when the cloud population becomes unstable, and some drops grow at the expense of others. There are two mechanisms whereby a cloud microstructure may become unstable. The first is the direct collision and coalescence (sticking) of water droplets and may be important in any cloud. The second mechanism requires the interaction between water droplets and ice

crystals and is confined to those clouds whose tops extend to temperatures colder than 0°C.

From analysis of the aerodynamic forces it is found that very small
droplets cannot readily be made to collide. A small drop falling through
a cloud of still smaller droplets will collide with only a minute fraction
of the droplets in its path if its radius is less than about 18 μm. Therefore it is expected that clouds which contain negligible numbers of
drops larger than 18 μm will be relatively stable with respect to growth by
coalescence. Clouds with considerable numbers of larger drops may
develop precipitation.

When an ice crystal exists in the presence of a large number of
supercooled water droplets the situation is immediately unstable. The
equilibrium vapor pressure over ice is less than that over water at the
same temperature and consequently the ice crystal grows by diffusion of
vapor and the drops evaporate to compensate for this. The vapor transfer
depends on the difference in equilibrium vapor pressure of water and ice
and is most efficient at about -15°C.

Once the ice crystal has grown by diffusion to a size appreciably larger
than the water droplets, it begins to fall relative to them and collisions
become possible. If the collisions are mainly with other ice crystals
snowflakes form; if water droplets are collected graupel or hail may
form. Once the particle falls below the 0°C level melting can occur, and
the particle may emerge from cloud base as a raindrop indistinguishable

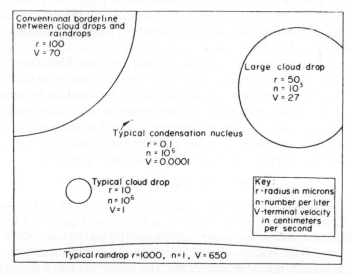

FIG. 6.1. Comparative sizes, concentrations, and terminal fall velocities of
some of the particles included in cloud and precipitation processes. (From
McDonald, 1958.)

from one formed by coalescence. In cold weather, or when large hailstones are formed, the particle may of course reach the ground unmelted.

The particles of interest in cloud physics have a wide range of size, concentration, and fall velocity. Figure 6.1 compares these properties for the particles important for condensation and precipitation processes. Noteworthy are the vast differences in size between a typical condensation nucleus and a cloud drop, and between cloud drop and raindrop. To account for the production of natural rain the growth processes must be fast enough to get from condensation nucleus to raindrop in about 20 min.

Nucleation of liquid water in water vapor

The problem of nucleation may be stated as follows: How readily can chance collisions and aggregations of water molecules lead to the formation of an embryonic droplet that will be stable and continue to exist under given environmental conditions?

The embryonic droplet will be stable if its size exceeds a certain critical value. On the average, droplets larger than the critical size grow while those that are smaller decay. What determines the critical size is the balance between the opposing rates of growth and decay. These rates, in turn, depend upon whether the droplet forms in free space (homogeneous nucleation) or in contact with another body (heterogeneous nucleation). For the homogeneous nucleation of pure water, the growth rate depends upon the partial pressure of water vapor in the surroundings because this determines the rate at which water molecules impinge upon the droplet. The decay process, evaporation, depends strongly upon the temperature of the droplet and its surface tension. Molecules at the surface of the drop must obtain sufficient energy to overcome the binding forces if they are to escape. If equilibrium is established between the liquid and its vapor, the rates of condensation and evaporation are exactly balanced and the vapor pressure is equal to the equilibrium or saturation vapor pressure. But the equilibrium vapor pressure over a droplet's surface depends upon its curvature and is given by

$$e_s(r) = e_s(\infty) \exp (2\sigma/rR_v\varrho_L T), \qquad (6.1)$$

where $e_s(r)$ is the saturation vapor pressure over the surface of a spherical droplet of radius r with surface tension σ and density ϱ_L at temperature T. R_v is the gas constant for water vapor and $e_s(\infty)$ is the saturation vapor pressure over bulk water—the quantity most readily measured. Note that as the size of the drop decreases, the vapor pressure required for equilibrium with it increases. This equation was first derived by William

Thomson (1870) (later Lord Kelvin), in slightly different form, to explain the rise of liquids in capillary tubes.

The surface tension σ is the free energy per unit surface area of the liquid. It can be defined as the work per unit area required to extend the surface of the liquid at constant temperature. Because work is the product of force and distance, work per unit area has units of force per unit length. The surface tension of water is approximately 7.5×10^{-2} N/m for meteorological temperatures. (See problem 6.9.)

The net rate of growth of a droplet of radius r is proportional to the difference $e - e_s(r)$, where e denotes the actual ambient vapor pressure. Drops with radii such that $e - e_s(r) < 0$ tend to decay while those with radii such that $e - e_s(r) > 0$ tend to grow. The critical size r_c is defined as the radius for which $e - e_s(r_c) = 0$, or $e = e_s(\infty) \exp(2\sigma/r_c \varrho_L T)$. Hence

$$r_c = \frac{2\sigma}{R_v \varrho_L T \ln S}, \qquad (6.2)$$

where $S = e/e_s(\infty)$ denotes the *saturation ratio*. For a droplet to be stable which formed by chance collisions among water molecules, it must grow to a radius larger than r_c. The following table gives values of the critical droplet radius and the number of molecules in the critical droplet for various saturation ratios at 0°C.

The table shows that high supersaturations are required for very small droplets to be stable. For example, when the supersaturation is 1%, corresponding to $S = 1.01$, droplets with radii smaller than $0.121\,\mu$m are unstable and will tend to evaporate.

In homogeneous nucleation, droplets of critical size are formed by random collisions of water molecules. If these droplets capture another molecule, they become supercritical; that is, with increasing size, $e_s(r)$ decreases and the rate of growth (which is proportional to $e - e_s(r)$)

TABLE 6.1. *Radii and Number of Molecules in Droplets of Pure Water in Equilibrium with the Vapor at 0°C*

Saturation ratio S	Critical radius $r_c\,(\mu\text{m})$	Number of molecules n
1	∞	∞
1.01	1.208×10^{-1}	2.468×10^8
1.10	1.261×10^{-2}	2.807×10^5
1.5	2.964×10^{-3}	3.645×10^3
2	1.734×10^{-3}	730
3	1.094×10^{-3}	183
4	8.671×10^{-4}	91
5	7.468×10^{-4}	58
10	5.221×10^{-4}	20

increases. Supercritical droplets therefore grow spontaneously. The rate of nucleation is simply the rate at which supercritical droplets are formed and is given by the product of the concentration of critical droplets and the rate at which a critical droplet gains another molecule and becomes supercritical. From statistical thermodynamics the nucleation rate per unit volume is determined to be given approximately by

$$J = 4\pi r_c^2 \frac{e}{\sqrt{2\pi mkT}} Zn \exp\left(-\frac{4\pi r_c^2 \sigma}{3kT}\right), \qquad (6.3)$$

where m is the mass of a water molecule, k is Boltzmann's constant, and n is the number density of vapor molecules. The factor Z denotes the Zeldovich or non-equilibrium factor, which has a numerical value in the order of 10^{-2} when all quantities are measured in CGS units. By substituting (6.2) for r_c in (6.3), J is expressed as a function of S and T. An important characteristic of this relation is that, for a given temperature, the nucleation rate increases from undetectably small values to extremely large values over a very narrow range of S. By convention, the threshold defining a significant rate of homogeneous nucleation is taken to be $1 \text{ cm}^{-3} \text{ s}^{-1}$. Events occurring at this rate are readily observable in experiments. The value of S corresponding to $J = 1 \text{ cm}^{-3} \text{ s}^{-1}$ is called the critical saturation ratio S_c. Theory and experimental data (Miller *et al.*, 1983) show that S_c decreases with increasing temperature, having a value of approximately 4.3 at 273 K, 6.3 at 250 K, and 3.5 at 290 K. Such large saturation ratios are never observed in the atmosphere, where supersaturations rarely exceed 1 or 2%. Homogeneous nucleation of liquid water from its vapor is not possible in these conditions.

In the atmosphere cloud droplets form on the aerosols called condensation nuclei or hygroscopic nuclei. The rate of droplet formation is determined by the number of these nuclei present, and not by collision statistics. In general, aerosols can be classified according to their affinity for water as hygroscopic, neutral, or hydrophobic. Nucleation on a neutral aerosol requires about the same supersaturation as homogeneous nucleation. On a hydrophobic aerosol particle, which resists wetting, nucleation is more difficult and requires even higher supersaturations. But for the hygroscopic particles, which are soluble and have an affinity for water, the supersaturation required for droplet formation can be much less than its value for homogeneous nucleation.

A nonvolatile dissolved substance tends to lower the equilibrium vapor pressure of a liquid. Very roughly, the effect may be thought of as arising from the fact that, when solute is added to a liquid, some of the liquid molecules that were in the surface layer are replaced by solute molecules. If the vapor pressure of the solute is less than that of the solvent, the vapor pressure is reduced in proportion to the amount of solute present. This effect can drastically lower the equilibrium vapor pressure over a droplet; the result is that a *solution droplet* can be in equilibrium with an environment at much lower supersaturation than a pure water droplet of the same size.

For a plane water surface the reduction in vapor pressure due to the presence of a nonvolatile solute may be expressed

$$\frac{e'}{e_s(\infty)} = \frac{n_0}{n + n_0},$$

where e' is the equilibrium vapor pressure over a solution consisting of n_0 molecules of water and n molecules of solute. This is known as Raoult's Law. For dilute solutions, $n \ll n_0$ and

$$\frac{e'}{e_s(\infty)} = 1 - \frac{n}{n_0}. \tag{6.4}$$

For solutions in which the dissolved molecules are dissociated, (6.4) must be modified by multiplying n by the factor i, the degree of ionic dissociation. Low (1969) explained that the factor i, which is called the van't Hoff factor, may be determined from the coefficient of ionic activity, a more fundamental quantity for which abundant experimental data are available. He tabulated values of i over a range of concentrations for eight electrolytes, including sodium chloride and ammonium sulfate, which are important as cloud nuclei. For both, $i \approx 2$ appears to be a reasonable approximation to use in calculations in the absence of more precise information.

The number of effective ions in a solute of mass M is given by

$$n = iN_0M/m_s,$$

where N_0 is Avogadro's number (the number of molecules per mole) and m_s is the molecular weight of the solute. The number of water molecules in mass m may likewise be written

$$n_0 = N_0m/m_v.$$

Writing for the mass of water $m = \frac{4}{3}\pi r^3\varrho_L$, we note that (6.4) can now be expressed as

$$\frac{e'}{e_s(\infty)} = 1 - b/r^3, \tag{6.5}$$

where $b = 3im_vM/4\pi\varrho_Lm_s$.

Kelvin's equation (6.1) and the solution effect are combined to give for the equilibrium vapor-pressure $e_s'(r)$ of a solution droplet

$$\frac{e_s'(r)}{e_s(\infty)} = \left[1 - \frac{b}{r^3}\right]e^{a/r},$$

where $a = 2\sigma/\varrho_L R_v T$. For r not too small, a good approximation to this equation is

$$\frac{e_s'(r)}{e_s(\infty)} = 1 + a/r - b/r^3. \tag{6.6}$$

In this approximate form, a/r may be thought of as a "curvature term" which expresses the increase in saturation ratio over a droplet as compared to a plane surface. The term b/r^3 may be called the "solution term", for it shows the reduction in vapor pressure due to the presence of a dissolved substance. Numerically, $a \approx 3.3 \times 10^{-5}/T$ (cm) and $b \approx 4.3iM/m_s$ (cm^3). For given values of T, M, and m_s, (6.6) describes the dependence of saturation ratio on the size of a solution droplet. The resultant curve is called a Köhler curve, an example of which is illustrated in Fig. 6.2.

The curve shows that the solution effect dominates when the radius is small, so that a very small solution droplet is in equilibrium with the vapor at relative humidities less than 100%. If the relative humidity is increased a small amount, the droplet will grow until it reaches equilibrium once again. This process of increasing the ambient humidity and allowing the droplet to grow to equilibrium size can be continued up to the relative humidity of 100% and slightly beyond. Finally the critical saturation ratio S^* is reached that corresponds to the peak of the Köhler curve—in this example a supersaturation of 0.6%, corresponding to a

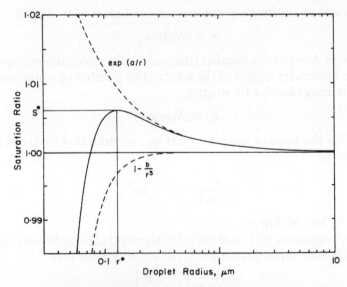

FIG. 6.2. Equilibrium saturation ratio of a solution droplet formed on an ammonium sulfate condensation nucleus of mass 10^{-16} g.

TABLE 6.2. *Values of* r* *and* (S* − 1) *as Functions of Nucleus Mass and Radius, assuming NaCl Spheres at a Temperature of 273°K*

Mass of dissolved salt (g)	r_s (µm)	r^* (µm)	$(S^* − 1)$ (%)
10^{-16}	0.0223	0.19	0.42
10^{-15}	0.0479	0.61	0.13
10^{-14}	0.103	1.9	0.042
10^{-13}	0.223	6.1	0.013
10^{-12}	0.479	19	0.0042

critical radius r^* of 0.13 µm. Up to this point the relative humidity had to be increased for the droplet to grow. But notice that if the relative humidity slightly exceeds S^*, enabling the droplet to grow beyond r^*, its equilibrium saturation ratio falls below S^*. Consequently the vapor will diffuse to the droplet and it will continue to grow without the need for further increase in the ambient saturation ratio. Up to size r^*, which for most droplets is between 0.1 and 1 µm, the droplet is in a stable equilibrium with its environment: any change in saturation ratio causes the droplet either to grow or evaporate until again reaching equilibrium. Beyond r^* the equilibrium is unstable. Any change in saturation ratio causes the droplet to grow or evaporate but to deviate further from the size for equilibrium. When smaller than r^* and in stable equilibrium, a droplet is called a haze particle. A condensation nucleus is said to be "activated" when the droplet formed around it reaches the size r^*. Once the droplet exceeds this size it will continue to grow to cloud droplet size if the ambient saturation ratio remains at a value above the equilibrium curve. In actual clouds the growth does not continue indefinitely because many droplets are present, which compete for the available vapor and tend to lower the saturation ratio once the condensation becomes more rapid than the production of supersaturation.

The critical values of radius and saturation ratio can be derived from the approximate expression (6.6) and are given by

$$r^* = \sqrt{3b/a} \tag{6.7}$$

$$S^* = 1 + \sqrt{4a^3/27b}. \tag{6.8}$$

Table 6.2 gives examples of critical radii and supersaturations for droplets formed on sodium chloride nuclei.

Atmospheric condensation nuclei

Aerosol particles, some of which are hygroscopic, are formed by the condensation of gases or by the disintegration of liquid or solid material. Particles formed by condensation are usually spherical. Others may be crystals, fibers, agglomerates, or irregular fragments. For convenience, all are often described in terms of an equivalent spherical diameter, which is the diameter of a sphere having the same volume as the aerosol particle. The particles range in size from about 10^{-3} μm diameter for small clusters of a few molecules to more than 10 μm for large salt, dust, and combustion particles.

According to Brock (1972), about 75% of the total mass of aerosol material in the atmosphere is accounted for by natural and anthropogenic primary sources such as wind-generated dust (20%), sea spray (40%), forest fires (10%), and combustion and other industrial operations (5%). The remaining 25% is attributed to secondary sources which involve conversion of gaseous constituents of the atmosphere to small particles by photochemical and other chemical processes. Chief among the gases that react to form particulates are SO_2, NO_2, olefins, and NH_3. Regardless of their mechanism of introduction, atmospheric aerosols continually undergo many chemical and physical transformations, including coagulation, condensation, scavenging, washout, sedimentation, dispersion, and mixing.

Any given sample of aerosols usually contains particles with a broad range of size. By convention, particles in the range $0.2\,\mu$m $< D < 2\,\mu$m are called large aerosols and those larger than 2 μm are called giant aerosols. Particles smaller than 0.2 μm in diameter, which include the overwhelming majority of atmospheric aerosols, are called Aitken particles, after the Scottish physicist of the last century who developed instruments for observing aerosol particles. These categories of particle size were introduced when the practice was to describe aerosol size by radius instead of diameter.

In an Aitken particle counter, versions of which are still in use today, a sample of air is drawn into a chamber, humidified, and subjected to a sudden expansion. The expansion causes a supersaturation typically of a hundred per cent of more, activating most of the aerosol particles to form a cloud of small water droplets. In modern instruments the concentration of droplets is estimated automatically from a measurement of the attenuation of a beam of light across the cloud chamber. Only a very small fraction of the aerosols activated in an Aitken counter are important in natural clouds, because supersaturations in the free atmosphere are thought only rarely to exceed 1%.

Aerosol concentrations—expressed as the number of particles per unit volume of space—are highly variable in time and position. Concentrations are greatest near the ground and near obvious sources, such as cities, industrial sites, fires, or active volcanoes. A rather high value is 10^5 cm^{-3}, which may be taken as a typical concentration in polluted air. Values can be one or two orders of magnitude higher close to the sources of aerosol particles, and several orders of magnitude lower in clean air.

Aerosol mass concentrations, even in heavily polluted air, are usually less than $1000\,\mu\text{g/m}^3$.

The size distribution of particles in an aerosol sample may be described by the distribution function $n_v(v)$, with the property that $n_v(v)dv$ is the number of particles per unit volume of air whose sizes are between v and $v + dv$ in volume. With sizes expressed as the diameter of equivolume spheres, the appropriate distribution function is $n_d(D)$, with the property that $n_d(D)dD$ is the number of particles per unit volume whose diameters are between D and $D + dD$. Because the number of particles in a given size interval must be the same regardless of the scale on which the size is measured, the two size distribution functions are related by

$$n_d(D)dD = n_v(v)dv. \tag{6.9}$$

The relation between volume and equivalent spherical diameter is

$$v = \pi D^3/6.$$

Therefore the particle size distribution with diameter the distributed variable may be written in terms of the basic distribution function n_v as

$$n_d(D) = n_v(v)\frac{dv}{dD} = \frac{\pi}{2}D^2 n_v\left(\frac{\pi D^3}{6}\right). \tag{6.10}$$

Similarly, a particle size distribution can be defined with surface area as the distributed variable.

The number of particles per unit volume with diameter smaller than D is

$$N(D) = \int_0^D n_d(D')dD', \tag{6.11}$$

where D' is a dummy variable of integration that stands for diameter. The function $N(D)$ is called the cumulative distribution of particle diameter. Clearly

$$n_d(D) = \frac{dN}{dD}. \tag{6.12}$$

Because of the wide range over which particle sizes can vary, it is convenient to use a logarithmic scale. One method is to plot $\log n_d$ versus $\log D$. Another is to define a distribution function $n_l(D)$ such that $n_l(D)d(\log D)$ is the number of particles per unit volume whose diameters are in the interval $d(\log D)$. In terms of the cumulative distribution function,

$$n_l(D) = \frac{dN(D)}{d(\log D)}. \tag{6.13}$$

From (6.12), it follows that

$$n_l(D) = D \ (\ln 10) \ \frac{dN}{dD} = D \ (\ln 10) \ n_d(D), \qquad (6.14)$$

which is the relation between the distribution function $n_l(D)$ for logarithmic increments of diameter and the function $n_d(D)$ for linear increments.

Plots of $\log n_l(D)$ versus $\log D$ for aerosol particles can sometimes be fit by straight lines over a limited range of diameter. This implies a power law dependence of n_l on D of the form

$$n_l(D) = cD^{-\beta}, \qquad (6.15)$$

where c is a constant and $-\beta$ is the slope of the straight line. $\beta = 3$ is often cited as typical over the diameter range from about 10^{-1} to $10\,\mu m$; aerosol populations having this form are said to follow a Junge distribution, after the atmospheric chemist Christian Junge, an authority on aerosols. From (6.14), it is seen that a power law of the form (6.15) for $n_l(D)$ corresponds to a power law for $n_d(D)$ given by

$$n_d(D) = \frac{c}{(\ln 10)} D^{-(\beta+1)}.$$

The differences among aerosol samples are emphasized by comparing the contributions of given size intervals to the total particulate volume or surface area rather than to the total number of particles. For example, the surface area of particles with diameters smaller than D is given by

$$S(D) = \int_0^D \pi D'^2 n_d(D') dD'. \qquad (6.16)$$

The contribution of particles in $d(\log D)$ to the surface area is

$$\frac{dS}{d(\log D)} = D \ (\ln 10) \ \frac{dS}{dD} = \pi D^2 \ \frac{dN}{d(\log D)} = \pi D^2 n_l(D). \quad (6.17)$$

Similarly, the volume of particles with diameters smaller than D is

$$V(D) = \int_0^D \frac{\pi}{6} D'^3 n_d(D') dD' \qquad (6.18)$$

and the contribution of particles in $d(\log D)$ to the volume is

$$\frac{dV}{d(\log D)} = \frac{\pi}{6} D^3 n_l(D). \qquad (6.19)$$

Figure 6.3 shows the size distributions of aerosols collected by airplane over a north-central region of the United States (Miles City, Montana).

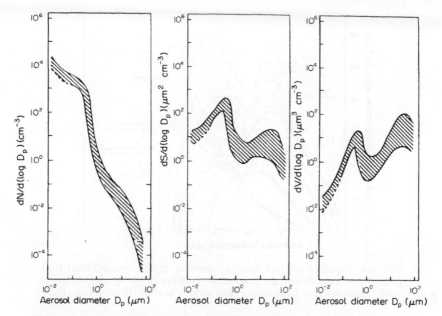

FIG. 6.3. Size distributions of aerosol particles measured by airplane in the lower troposphere. The same data are plotted three ways. The envelopes indicate the variability of the observations. (Adapted from Hobbs *et al.*, 1985.)

The hatching defines an envelope that includes many individual samples. The data are plotted three ways: as distributions of particle number $dN/d(\log D)$, surface area $dS/d(\log D)$, and volume $dV/d(\log D)$. The Junge distribution or any other power law provides only a crude fit to the data and is unable to capture the inflection points, maxima, and minima, which are conspicuous features of the volume and surface area distributions. The maximum in these distributions between 0.1 and 1 μm is called the accumulation mode; that between 10 and 20 μm is called the coarse particle mode. Not evident here is another mode near $10^{-2}\,\mu$m occasionally observed near industrial centers and other sources of combustion, called the nucleation mode.

The accumulation mode is explained by the tendency of particles smaller than 0.1 μm to collide with one another as a result of Brownian motion and to clump together. The large particle mode is dominated by dust, combustion products, and sea spray, and depends on surface wind speed and distance from the sources of the particulates. An upper limit to the sizes is established by sedimentation, which preferentially removes the larger particles from the distribution.

The most important removal process for aerosol particles is precipitation. Some particles serve as centers for cloud condensation; others are

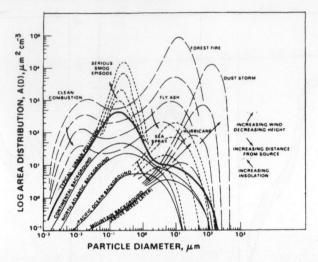

FIG. 6.4. Idealized aerosol spectra, showing typical ground-level background distributions and the general dependence of the spectra on height,
wind speed, distance from source, and surface heating. (From Slinn, 1975.)

captured by droplets as they continue to grow by diffusion from the
vapor, or are swept out by falling precipitation particles. All such
particles are lost from the aerosol population, although new aerosols are
created as residue from evaporating cloud and precipitation particles.

Figure 6.4 is a schematic illustration of aerosol particle size distributions that typify different background and extreme conditions. The
coagulation mode between 0.1 and 1 μm is evident in most of the
distributions representative of conditions near the ground. The coarse
particle mode (diameter $>$ 10 μm) is pronounced in samples that are
influenced by nearby combustion sources or high winds, which stir up
dust from the ground.

From the entire aerosol population, the particles thought to be most
important in natural cloud formation are those in the accumulation
mode, which corresponds approximately to the size interval of the
so-called large aerosols. Particles smaller than about 0.1 μm, even if
hygroscopic, would require higher saturation ratios than those thought
to exist in the free atmosphere to be activated. Particles larger than 4 or
5 μm are much fewer in number (except near sources) but may serve as
nuclei for some of the cloud droplets that are much larger than average
size.

The chemical composition of cloud-forming nuclei is difficult to infer
because they comprise such a small fraction of the total atmospheric
aerosol population. Evidence indicates that sea salt is the predominant
constituent of condensation nuclei in the size range above 1 μm. The

main component of the important range between 0.1 and 1 μm diameter is thought to be sulfate, which may be present either as sulfuric acid or combined as a salt such as ammonium sulfate.

A useful way to describe the cloud forming propensity of an aerosol population is by its *activity spectrum*, which is the number of particles per unit volume that are activated to become cloud droplets, expressed as a function of the supersaturation. Such spectra are measured using cloud chambers in which slight supersaturations can be achieved and accurately controlled. An air sample is brought into the chamber and the super-saturation is fixed at a low value, typically a few tenths of a percent. By optical means the number of nuclei that grow to activation size are observed and counted. Counts are taken at steps of increasing super-saturation, usually over the range from about 0.3% to 1%. The nuclei activated in this way are called cloud condensation nuclei (CCN). They are the subset of the total aerosol population that can account for the formation of natural clouds. The nucleus counts may often be approxi-mated by the power-law relation

$$N_c = Cs^k, \tag{6.20}$$

where $s = (S - 1) \times 100\%$ is the percent supersaturation, N_c is the number of nuclei per unit volume activated at supersaturations less than s, and C and k are parameters that depend on the airmass type. Typical ranges are, for maritime air, $C = 30$ to 300 cm^{-3}, $k = 0.3$ to 1.0; for continental air, $C = 300$ to 3000 cm^{-3}, $k = 0.2$ to 2.0.

Condensation nuclei of some sort are always present in the atmosphere in ample numbers: clouds form whenever there are vertical air motions and sufficient moisture. However, in some marginal cases precipitation is more likely to occur when the nucleus population consists of a few large particles instead of many small particles.

Some of the ways of measuring atmospheric nuclei are inappropriate for the needs of cloud physics. For example, Aitken nucleus counts, as obtained with an expansion chamber counter, are essentially counts of *all* the condensation nuclei in an atmospheric sample. Supersaturations of several hundred per cent may be created in such instruments, activating nearly all the nuclei present. In the natural atmosphere, the only nuclei that enter cloud forming processes are the CCN that are activated at supersaturations less than a few per cent.

Assuming an activity spectrum of the form (6.20), Twomey (1959) showed that the droplet concentration N formed in an updraft of speed U can be expressed in terms of U, C, and k. For k between 0.4 and 1.0, his result may be approximated by

$$N \approx 0.88C^{2/(k+2)}[7 \times 10^{-2}U^{3/2}]^{k/(k+2)} \tag{6.21}$$

where N is in cm^{-3} and U is in cm/s. Twomey also obtained an expression for the peak supersaturation in the updraft, which may be approximated by

$$s_{max} \approx 3.6[(1.6 \times 10^{-3}U^{3/2})/C]^{1/(k+2)}. \tag{6.22}$$

Comparisons of activity spectra with observed droplet spectra provide experimental confirmation of the close relation between the nucleus population and the resulting cloud droplets. The development of a cloud after its formative stage, and in particular the amount and character of precipitation produced, is controlled more by large-scale phenomena, such as updraft speed and moisture supply, than by its microphysical structure. But to some extent the microstructure determines how susceptible a cloud is to producing precipitation, and how much time is required for the precipitation to form.

Problems

6.1. The following table, from Friedlander (1977, p. 9), gives the times required for the concentration of aerosol particles of 0.1 μm diameter to decrease by coagulation. Listed are the times for the concentration, initially equal to N_0, to fall to one-tenth that value.

$N_0 (cm^{-3})$	10^8	10^7	10^6	10^5
$t_{1/10}$	2 min	20 min	3.5 h	35 h

Show that these data are consistent with the following formula for the reduction of aerosol concentration by coagulation:

$$\frac{dN}{dt} = -kN^2.$$

Solve for the value of k from the data.

6.2. A haze droplet is activated to form a cloud droplet when the ambient saturation ratio exceeds the critical value S^* and the droplet radius exceeds r^*. At this point the original condensation nucleus may be assumed to be completely dissolved in the droplet. Prove that, for a given hygroscopic material and a fixed temperature, the concentration of the solution when $r = r^*$ is stronger for small condensation nuclei than for large ones. Evaluate the concentration, in units of mass of salt per unit mass of water, for sodium chloride nuclei ranging from 10^{-19} to 10^{-14} g.

6.3. Consider two cloud droplets, one formed on a sodium chloride nucleus of mass 10^{-16} g, the other on an ammonium sulfate nucleus of the same mass. Calculate the critical size r^* and saturation ratio S^* for each droplet, assuming a temperature of 280 K. For any radius r, let $S_1(r)$ denote the equilibrium saturation ratio of the droplet formed on sodium chloride and $S_2(r)$ denote the equilibrium saturation ratio of the droplet formed on ammonium sulfate. For $r > r_2^*$, the activation size of the ammonium sulfate droplet, show that $(S_2 - S_1)$ and S_2/S_1 decrease monotonically with increasing r to the limits 0 and 1, respectively.

6.4. A given population of aerosol particles may be approximated as a Junge distribution extending from 0.1 μm to 10 μm in diameter. That is,

$$n_l(D) = \begin{cases} cD^{-3}, & 0.1\ \mu m \le D \le 10\ \mu m \\ 0, & \text{otherwise}. \end{cases}$$

If the total volume of these aerosols is 10^{-9} cm^3 per cm^3 of air, solve for the number density of aerosols in cm^{-3} and their total surface area in cm^2/cm^3.

6.5. In an Aitken nucleus counter, an air sample is drawn into a chamber, humidified, and suddenly expanded. The expansion creates a high value of supersaturation, activating the nuclei with thresholds below this value. Suppose the air is initially saturated at temperature T_0 and expands adiabatically into a volume Ω times larger than the initial volume. Neglecting any condensation during the expansion process, solve for the supersaturation as a function of the initial temperature T_0 and the expansion ratio Ω. Show that the supersaturation is approximately 220% for an expansion ratio of 1.2 and an initial temperature of 15°C.

6.6. Thermal gradient diffusion chambers are used to establish an experimental environment with a known and precisely adjustable level of supersaturation for studies in heterogeneous nucleation of the liquid phase. Such a chamber is indicated schematically below.

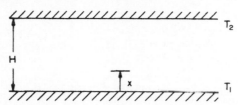

The top and bottom surfaces are kept moist at temperatures T_2 and T_1, respectively, with $T_2 > T_1$. The temperature difference across the chamber is thus $T_2 - T_1$ and the vapor density difference is $\varrho_v(T_2) - \varrho_v(T_1)$, where $\varrho_v(T)$ denotes the saturation vapor density at temperature T. In a steady state, heat and vapor flow from the warm upper surface to the cool bottom surface and the profiles of temperature and vapor density are linear. That is, the temperature gradient is $(T_2 - T_1)/H$ and the vapor density gradient is $[\varrho_v(T_2) - \varrho_v(T_1)]/H$, and both are constant for $0 < x < H$.

(a) In the steady state, prove that the chamber is supersaturated at all positions x such that $0 < x < H$.
(b) Show that if the temperature difference is small $[(T_2 - T_1) \ll T_1]$, the maximum supersaturation is located approximately midway between the upper and lower surfaces.
(c) Assuming $T_1 = 273$ K, solve for the maximum supersaturation as a function of T_2 over the range 273 K $< T_2 <$ 283 K.

6.7. An air sample contains aerosol particles of sodium chloride with sizes ranging from D_0 to D_{max} and distributed according to the Junge form of problem 6.4. Nucleus counts are taken in a thermal-gradient diffusion chamber by observing the number of droplets that are activated as a function of the supersaturation. If D_{max} is large, show that the number of activated droplets is related to supersaturation by $N \propto s^2$.

6.8. The terminal velocity (settling speed) of a spherical aerosol particle of diameter D is given to good approximation by Stokes' Law,

$$u_p = \frac{\varrho_p D^2 g}{18\mu},$$

where ϱ_p is the density of the particle and μ is the dynamic viscosity of the air (see Table 7.1, p. 103). Derive an expression for the rate of sedimentation of aerosol material from the air in units of mass per unit area and time. For the aerosol sample of problem 6.4, assume a density of 2 g/cm³ and calculate the sedimentation rate in units of μg m^{-2} 24 h^{-1}.

6.9. Over the temperature range from $-20°C$ to $+20°C$ laboratory data indicate that the surface tension of water varies inversely with temperature. The data can be accurately approximated by the formula $\sigma = c_1 T + c_2$, where $c_1 = -1.55 \times 10^{-4}$ N m^{-1} K^{-1} and $c_2 = 0.118$ N m^{-1}, with T in K. Use this expression for σ in (6.1) and solve for the droplet size that must be exceeded if $e_s(r)$ is to increase with increasing temperature. Show that all embryonic water droplets exceed this size.

6.10. Cloud droplets begin to form on a population of condensation nuclei as the saturation ratio is gradually increased, with temperature held constant. The nuclei have the same chemical composition but a broad range of sizes. Droplets are activated with radii $r^* = 0.5\,\mu$m when the supersaturation reaches 0.15%. Droplets continue to be activated as the supersaturation is raised to 1%. Solve for r^* corresponding to a supersaturation of 1%. Also solve for the temperature.

7

Droplet Growth by Condensation

Diffusional growth of a droplet

It was shown in Chapter 6 that a critical size r^* and saturation ratio S^* must be exceeded for a small solution droplet to become a cloud droplet. Before and after the droplet reaches the critical size, it grows by diffusion of water molecules from the vapor onto its surface. The rate of diffusional growth of a single droplet is analyzed in this section. Later we shall consider the more realistic situation of many droplets simultaneously growing and competing for the available moisture.

The droplet has radius r and is located in a vapor field with the concentration of vapor molecules at distance R from the droplet center denoted by $n(R)$. The vapor field could as well be described in terms of the vapor density or absolute humidity $\varrho_v(R)$, where $\varrho_v = nm_0$ and m_0 denotes the mass of one water molecule. Isotropy is assumed so that $n(R)$ or $\varrho_v(R)$ does not depend on the direction outwards from the droplet. At any point in the vapor field the concentration of molecules is assumed to satisfy the diffusion equation

$$\frac{\partial n}{\partial t} = D\nabla^2 n, \tag{7.1}$$

where D is the molecular diffusion coefficient. Furthermore, steady-state or "stationary" conditions are assumed, so that $\partial n/\partial t = 0$. Then (7.1) becomes

$$\nabla^2 n(R) = 0 = \frac{1}{R^2}\frac{\partial}{\partial R}\left(R^2\frac{\partial n}{\partial R}\right), \tag{7.2}$$

with the general solution

$$n(R) = C_1 - C_2/R. \tag{7.3}$$

Boundary conditions are as follows:

as $R \to \infty$, $n \to n_\infty$, the "ambient" or undisturbed value of vapor concentration;

as $R \to r$, $n \to n_r$, the vapor concentration at the droplet's surface.

99

The solution satisfying these conditions is

$$n(R) = n_\infty - \frac{r}{R}(n_\infty - n_r).$$ (7.4)

The flux of molecules onto the surface of the droplet is equal to $D(\partial n/\partial R)_{R=r}$. Consequently the rate of mass increase is given by

$$\frac{dm}{dt} = 4\pi r^2 D \left(\frac{\partial n}{\partial R} \right)_{R=r} m_0.$$ (7.5)

Combining (7.4) and (7.5) gives

$$\frac{dm}{dt} = 4\pi r D(n_\infty - n_r)m_0.$$ (7.6)

In terms of the vapor density, this result may be written

$$\frac{dm}{dt} = 4\pi r D(\varrho_v - \varrho_{vr}),$$ (7.7)

where ϱ_v is the ambient vapor density and ϱ_{vr} is the vapor density at the droplet's surface.

This is the diffusional growth equation for an isolated droplet at rest in a vapor field. It shows that the droplet grows if $\varrho_v > \varrho_{vr}$ and evaporates if $\varrho_v < \varrho_{vr}$. Ordinarily, ϱ_v may be determined from given environmental conditions. ϱ_{vr} depends on the droplet size, chemical composition, and temperature. The temperature is usually not the same as the ambient temperature, and must be determined by considering the heat transfer between the droplet and its surroundings.

Associated with condensation is the release of latent heat, which tends to raise the droplet temperature above the ambient value. The diffusion of heat away from the droplet is given by an equation analogous to (7.7):

$$\frac{dQ}{dt} = 4\pi r K(T_r - T)$$ (7.8)

where T is ambient temperature, T_r is the temperature at the surface of the droplet, and K is the coefficient of thermal conductivity of air.

Equations (7.7) and (7.8) were first derived by James Clerk Maxwell in 1877, in slightly different form, in an article in *Encyclopaedia Britannica* on the theory of the wet-bulb thermometer (Maxwell, 1890). The theory of the steady-state growth of a spherical drop at rest in a vapor field regarded as a continuum is therefore often called the Maxwell theory.

From (7.7) and (7.8), the rate of change of temperature at the droplet's surface is given by

$$\frac{4}{3} \pi r^3 \varrho_L c \frac{dT_r}{dt} = L \frac{dm}{dt} - \frac{dQ}{dt} \tag{7.9}$$

where ϱ_L is the density of water and c is its specific heat capacity. For the assumed steady-state growth process, $dT_r/dt = 0$, so that (7.9), when set equal to zero, leads to a balance condition that must be satisfied by the temperature and vapor density fields,

$$\frac{\varrho_v - \varrho_{vr}}{T_r - T} = \frac{K}{LD}. \tag{7.10}$$

In this equation the ambient conditions are described by ϱ_v and T, and the ratio K/LD depends weakly on temperature and pressure. Ordinarily the drop temperature T_r and the vapor density at its surface ϱ_{vr} are unknown. From (6.6) and the equation of state for water vapor, these quantities are related by

$$\varrho_{vr} = e'_s(r)/R_v T_r = \left(1 + \frac{a}{r} - \frac{b}{r^3}\right) e_s(T_r)/R_v T_r, \tag{7.11}$$

where $e_s(T_r)$ is the equilibrium vapor pressure over a plane water surface at temperature T_r, and is given by the Clausius–Clapeyron equation. Equations (7.10) and (7.11) comprise a simultaneous system which can be solved numerically for T_r and ϱ_{vr} to permit evaluation of the rate of drop growth by condensation.

As an alternative to the numerical method of solution, Mason (1971) introduced a useful analytical approximation for calculating the rate of growth of a drop by condensation. In a field of saturated vapor, changes in vapor density are related to changes in temperature by

$$\frac{d\varrho_v}{\varrho_v} = \frac{L}{R_v} \frac{dT}{T^2} - \frac{dT}{T}. \tag{7.12}$$

Integrating this equation from temperature T_r to temperature T, and assuming $T/T_r \approx 1$, gives

$$\ln \frac{\varrho_{vs}}{\varrho_{vrs}} = (T - T_r)\left(\frac{L}{R_v T_r T} - \frac{1}{T_r}\right), \tag{7.13}$$

where the subscripts s indicate saturation vapor densities. Because $\varrho_{vs}/\varrho_{vrs}$ is close to unity, (7.13) leads to the approximate relation

$$\frac{\varrho_{vs} - \varrho_{vrs}}{\varrho_{vrs}} = \left(\frac{T - T_r}{T}\right)\left(\frac{L}{R_v T} - 1\right), \tag{7.14}$$

where the approximation has also been employed that $TT_r \approx T^2$. Substituting from (7.8) for $(T - T_r)$ gives

$$\frac{\varrho_{vs} - \varrho_{vrs}}{\varrho_{vrs}} = \left(1 - \frac{L}{R_v T}\right)\left(\frac{L}{4\pi r KT}\right)\frac{dm}{dt}. \tag{7.15}$$

From (7.7)

$$\frac{\varrho_v - \varrho_{vr}}{\varrho_{vr}} = (4\pi r D\varrho_{vr})^{-1}\frac{dm}{dt}. \tag{7.16}$$

Subtracting (7.15) from (7.16), and assuming that $\varrho_{vr} = \varrho_{vrs}$, leads eventually to the approximate result

$$r\frac{dr}{dt} = \frac{S - 1}{\left[\left(\dfrac{L}{R_v T} - 1\right)\dfrac{L\varrho_L}{KT} + \dfrac{\varrho_L R_v T}{De_s(T)}\right]} \equiv \frac{S - 1}{[F_k + F_d]}, \tag{7.17}$$

where $S = e/e_s(T)$ is the ambient saturation ratio. This form of the approximation neglects the solution and curvature effects on the drop's equilibrium vapor pressure. When these effects are included, the approximation becomes

$$r\frac{dr}{dt} = \frac{(S - 1) - \dfrac{a}{r} + \dfrac{b}{r^3}}{[F_k + F_d]}. \tag{7.18}$$

Here as in (7.17) F_k represents the thermodynamic term in the denominator that is associated with heat conduction; F_d is the term associated with vapor diffusion. In the term represented by F_k, it may be shown that $L/R_v T \gg 1$. As a close approximation, this term is sometimes written without the -1, which serves only as a small correction factor.

For typical values of a and b, it has been found that the droplet growth predicted by (7.18) is an excellent approximation to that obtained by simultaneous solution of (7.10) and (7.11).

The coefficients of diffusion and thermal conductivity vary with temperature and are tabulated in Table 7.1. The latent heat L and equilibrium vapor pressure e_s also depend on temperature, and are tabulated in Table 2.1 of Chapter 2. To a good approximation K is proportional to the dynamic viscosity μ of the air, which quantity is also included in the table. D is proportional to the kinematic viscosity, which equals μ/ϱ, with ϱ the air density. The dynamic viscosity depends only on temperature, and may be calculated from the approximate formula

$$\mu(T) = 1.72 \times 10^{-5}\left(\frac{393}{T + 120}\right)\left(\frac{T}{273}\right)^{3/2}$$

where T is in K and μ is in kg m^{-1} s^{-1}.

It is not possible to integrate the general equation (7.18) analytically. Given the temperature, pressure, saturation ratio, and mass and molecu-

TABLE 7.1. *Values of Dynamic Viscosity μ and Coefficient of Thermal Conductivity* K *of Air, and Coefficient of Diffusion* D *of Water Vapor in Air. (From Houghton, 1985)*

$T(°C)$	$\mu\,(\mathrm{kg\,m^{-1}s^{-1}})$	$K\,(J\,m^{-1}s^{-1}K^{-1})$	$D\,(m^2\,s^{-1})$
−40	1.512×10^{-5}	2.07×10^{-2}	1.62×10^{-5}
−30	1.564×10^{-5}	2.16×10^{-2}	1.76×10^{-5}
−20	1.616×10^{-5}	2.24×10^{-2}	1.91×10^{-5}
−10	1.667×10^{-5}	2.32×10^{-2}	2.06×10^{-5}
0	1.717×10^{-5}	2.40×10^{-2}	2.21×10^{-5}
10	1.766×10^{-5}	2.48×10^{-2}	2.36×10^{-5}
20	1.815×10^{-5}	2.55×10^{-2}	2.52×10^{-5}
30	1.862×10^{-5}	2.63×10^{-2}	2.69×10^{-5}

Note: The tabulated values of D are for a pressure of 100 kPa. Because D is proportional to μ/ϱ, it follows that D is inversely proportional to pressure for a given temperature. To obtain D for an arbitrary pressure p (kPa), the tabulated value must therefore be multiplied by $(100/p)$.

lar weight of the condensation nucleus, the equation can only be solved numerically or graphically to determine the droplet size as a function of time. Table 7.2 lists results for droplets growing on nuclei of sodium chloride at a supersaturation of 0.05%, $p = 90$ kPa, $T = 273$ K. A droplet that forms on a large condensation nucleus is seen to grow initially at a faster rate than droplets with small nuclei, but after a certain radius is reached (in this comparison, about $10\,\mu$m) the growth rates are about the same, regardless of nuclear mass. When a droplet becomes sufficiently large, the a/r and b/r^3 terms are negligible compared to $(S-1)$ and (7.17) is a good approximation. Then the droplet radius increases with time according to

$$r(t) = \sqrt{r_0^2 + 2\xi t}, \qquad (7.19)$$

TABLE 7.2. *Rate of Growth of Droplets by Condensation (initial radius 0.75 μm). (From Mason, 1971)*

Nuclear mass (g)	10^{-14}	10^{-13}	10^{-12}
Radius (μm)	Time (sec) to grow from initial radius 0.75 μm		
1	2.4	0.15	0.013
2	130	7.0	0.61
4	1,000	320	62
10	2,700	1,800	870
20	8,500	7,400	5,900
30	17,500	16,000	14,500
50	44,500	43,500	41,500

where

$$\xi = (S - 1)/[F_k + F_d]. \tag{7.20}$$

We may write $\xi = (S - 1) \cdot \xi_1$, where ξ_1 is a normalized condensation growth parameter defined by

$$\xi_1 = \frac{1}{F_k + F_d}.$$

ξ_1 is a function of temperature and pressure, and in the SI system has units of $m^2\,s^{-1}$. Because droplet size is usually measured in μm, it is more convenient to express ξ_1 in $\mu m^2\,s^{-1}$. Figure 7.1 shows the dependence of $\log_{10} \xi_1$, expressed in these units, on temperature and pressure. For example, at $p = 80$ kPa and $T = 0°C$, interpolation from the figure gives $\xi_1 = 10^{1.834} = 68.2\ \mu m^2\,s^{-1}$. The dashed lines in this figure represent the temperature and pressure along pseudoadiabats corresponding to $\theta_w = 0°C$ and $20°C$. These give an indication of the variation in the normalized growth parameter throughout the vertical extent of a cloud.

The parabolic form of (7.19) leads to a narrowing of the drop-size distribution as growth proceeds. For example, consider two droplets growing under the same conditions with initial radii of $r_1(0)$ and $r_2(0)$, with $r_2 > r_1$. Then (7.19) shows that the difference between the squares of the radii remains constant, so that at any time t,

$$r_2(t) - r_1(t) = \frac{r_2^2(0) - r_1^2(0)}{r_2(t) + r_1(t)},$$

and the difference in radii becomes smaller as they grow.

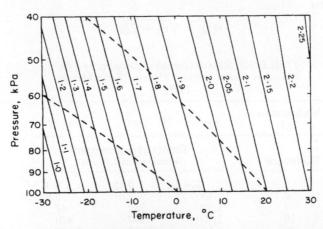

FIG. 7.1. Dependence of the growth parameter $\xi_1 = 1/[F_k + F_d]$ on temperature and pressure. Contours are plotted of the quantity $\log_{10} \xi_1$, with ξ_1 expressed in units of $\mu m^2/s$. Dashed lines represent pseudoadiabats corresponding to $\theta_w = 0°C$ and $20°C$.

TABLE 7.3. *Distance a Drop Falls before Evaporating, assuming Isothermal Atmosphere with* $T = 280\,K$, $S = 0.8$

Initial radius	Distance fallen
$1\,\mu m$	$2\,\mu m$
$3\,\mu m$	$0.17\,mm$
$10\,\mu m$	$2.1\,cm$
$30\,\mu m$	$1.69\,m$
$0.1\,mm$	$208\,m$
$0.15\,mm$	$1.05\,km$

The rate of evaporation of a droplet is also described by (7.18); in this case $S < 1$ and $(dr/dt) < 0$. By knowing the dependence of droplet fall speed on size, it is possible through (7.18) to solve for the distance a drop falls during the time required for it to evaporate completely. For drops smaller than about 50 μm in radius the terminal fall speed increases approximately with the square of the radius and the distance of fall for complete evaporation increases with the fourth power of radius. This approximation has been employed (and extended somewhat beyond its range of accuracy) to give the results in Table 7.3. A rapid increase of distance fallen with radius is evident, leading to a reasonable basis of discriminating between cloud droplets and raindrops. Raindrops are those large enough to reach the ground before evaporating. Cloud droplets are those small enough to evaporate soon after leaving the cloud. By convention the dividing line is drawn at $r = 0.1$ mm. drops larger than this size stand a good chance of reaching the ground and are called raindrops. The drops whose radii are near 0.1 mm are referred to as drizzle drops.

The growth of droplet populations

In natural clouds the droplets interact with their environment and with each other in many ways that affect the droplet sizes and concentration. Droplets grow by condensation if the environment has an excess of vapor over the equilibrium value, as would be produced by chilling in pseudoadiabatic ascent. They evaporate if dry environmental air in sufficient amount is mixed with the cloudy air. Sedimentation (gravitational settling) is important for the larger droplets and tends to remove them from the cloud. Coagulation of droplets (coalescence) to form larger ones appears to be insignificant for droplets smaller than 10 μm in radius, but becomes increasingly important as they grow beyond this size. A goal of cloud physics is to understand the processes that shape the droplet population.

In the early development of a cloud the droplets are too small for sedimentation or coalescence to be important. Condensation is the dominant growth process. Because the ambient saturation ratio controls this process, it is necessary to examine first the vapor budget of a developing cloud. We shall assume that vapor is provided by saturated air that is cooled in ascent and that it is lost by condensation on the growing droplets. The rate of change of the saturation ratio may be written

$$\frac{dS}{dt} = P - C, \tag{7.21}$$

where P denotes a production term and C a condensation term. More specifically,

$$\frac{dS}{dt} = Q_1 \frac{dz}{dt} - Q_2 \frac{d\chi}{dt}, \tag{7.22}$$

where dz/dt is the vertical air velocity and $d\chi/dt$ is the rate of condensation, measured in units of mass of condensate per mass of air per unit time. Q_1 and Q_2 are thermodynamic variables given by

$$Q_1 = \frac{1}{T} \left[\frac{\varepsilon L g}{R'c_p T} - \frac{g}{R'} \right] \tag{7.23}$$

$$Q_2 = \varrho \left[\frac{R'T}{\varepsilon e_s} + \frac{\varepsilon L^2}{p T c_p} \right]. \tag{7.24}$$

Physically, $Q_1(dz/dt)$ is the increase of supersaturation due to cooling in adiabatic ascent; $Q_2(d\chi/dt)$ is the decrease in supersaturation due to the condensation of vapor on droplets. Q_1 and Q_2 are plotted in Fig. 7.2 as functions of temperature.

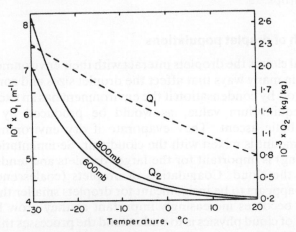

FIG. 7.2. The dependence of Q_1 and Q_2 on temperature. Q_2 depends also on pressure and is shown for 600 mb and 800 mb.

The derivation of (7.23) proceeds as follows. Assuming ascent with no condensation, (7.22) becomes

$$\frac{dS}{dt} = Q_1 \frac{dz}{dt}.$$ (7.25)

But

$$S = e/e_s,$$

so

$$\frac{dS}{dt} = \left(e_s \frac{de}{dt} - e \frac{de_s}{dt}\right)\Big/ e_s^2.$$ (7.26)

Introducing the mixing ratio w, and noting that w is constant under the assumption of no condensation, you can show that

$$\frac{de}{dt} = -\frac{eg}{R'T} \frac{dz}{dt}.$$ (7.27)

From the Clausius–Clapeyron equation,

$$\frac{de_s}{dt} = -\frac{Le_s}{R_v T^2} \frac{g}{c_p} \frac{dz}{dt}.$$ (7.28)

Using (7.26)–(7.28) in (7.25) leads to the result (7.23) for Q_1. The derivation of (7.24) follows along similar lines.

On the basis of (7.18) for the growth rate and (7.22) for the saturation ratio, it is possible to start with an assumed or measured distribution of condensation nuclei, specify an updraft velocity, and calculate the subsequent evolution of the droplet spectrum. Results of such a calculation are shown in Fig. 7.3. A weak updraft of 15 cm/s is assumed, with a moderate concentration of condensation nuclei at cloud base. The solid lines indicate the sizes of droplets growing on nuclei of sodium chloride having different masses, ranging from 3.7×10^{-18} to 9.3×10^{-10} g. The dashed envelope bounding the termination of these curves indicates the altitude reached by the different droplets during the simulation. The smaller ones move essentially with the air at 15 cm/s, but the larger ones fall relative to the air and do not rise to the same altitude. The dot-dashed line shows the variation with altitude of the percent supersaturation.

All droplets begin to grow as they ascend from cloud base. The supersaturation also increases, reaching a maximum of about 0.5% at an altitude slightly higher than 10 m above cloud base. The droplets that formed on the two smallest sizes of nuclei grow initially but evaporate after the supersaturation passes its maximum. This maximum is less than the critical supersaturation of either of the smallest nuclei, so that they remain as haze droplets and are not activated to become cloud drops.

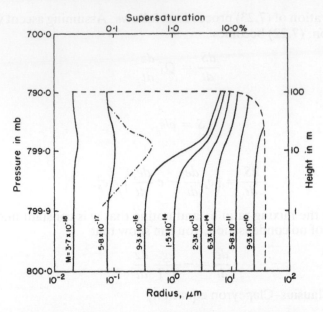

FIG. 7.3. Initial formation of cloud droplets and the variation of supersaturation above cloud base. (Adapted from Mordy, 1959.)

Droplets in the larger categories are activated, and experience rapid
growth during the period of high ambient supersaturation. As they
continue to grow their spread in size becomes narrower because of the
approximate parabolic form of the growth equation.

Figure 7.4 shows results of similar calculations, starting with a population of sodium chloride nuclei whose activity spectrum is approximately
of the form (6.20), with $C \approx 650 \, cm^{-3}$ and $k \approx 0.7$. Results are given for
two assumed constant updraft speeds, 0.5 and 2 m/s. The most important
cloud parameters are plotted as functions of altitude above the level
where $S = 1$ is reached. Notable are the sharp rise and gentle settling
down of supersaturation, the rapid increase of droplet concentration to a
steady value reached at the point of maximum supersaturation, and the
relative narrowness of the droplet size distribution, as measured by the
standard deviation. These and similar calculations show that the number
of cloud droplets produced in an updraft depends on the updraft speed
and the activity spectrum, and is generally consistent with Twomey's
equations (6.21) and (6.22). Calculated droplet spectra, however, are
often narrower and with fewer large droplets than observed droplet size
distributions.

A significant feature of all calculations of this kind is that the supersaturation reaches its peak within about 100 m of cloud base, above

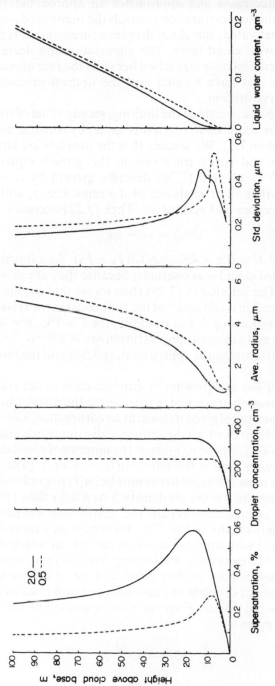

FIG. 7.4. Early development of cloud properties in air ascending at constant velocity of 0.5 m/s or 2 m/s.

which level it decreases and approaches an approximately constant value. Because the supersaturation controls the number of condensation nuclei that are activated, the cloud droplet concentration is thus determined in the lowest cloud layer. The supersaturation decreases to its steady value when a balance is reached between the rate of condensation on the droplets that have formed and the updraft-produced rate of increase of supersaturation.

We may derive an estimate of the limiting, steady value of supersaturation, and the time required to reach this value, by constructing a simple model (see problem 7.5). We assume that the droplets are large enough for the solution and curvature terms in the growth equation to be neglected. Then (7.19) and (7.20) describe growth by condensation. Assume further that all droplets are of the same size, r, and that their concentration is ν_0 per unit mass of air. Then (7.22) becomes

$$ds/dt = \omega - \eta s \qquad (7.29)$$

where $\omega = 100 Q_1 U$ and $\eta = 4\pi \varrho_L \nu_0 r Q_2 / (F_k + F_d)$. As a further approximation, we regard ω and η as constant, because they are slowly varying compared to s. The solution of (7.29) then shows that the limiting value of the supersaturation is ω/η and that the time constant or relaxation time of the supersaturation is η^{-1}. For example, for $T = 7°C$, $p = 80$ kPa, $r = 5\,\mu m$, $U = 5$ m/s, and a droplet concentration $\varrho\nu_0 = 300\,\mathrm{cm}^{-3}$, we find that the limiting supersaturation is approximately 0.5% and the time constant is 2 s.

When the droplets are growing by condensation under conditions of steady supersaturation, it is possible to solve for the droplet size distribution function at any time, given its form at an earlier time. Let us consider a sample of cloudy air in which the drop-size distribution is characterized by the function $\nu_0(r_0)$, where $\nu_0(r_0)dr_0$ is the number of cloud droplets per unit mass of air with radii in the interval $(r_0, r_0 + dr_0)$. (The number of droplets per unit mass is related to the number $n(r)$ per unit volume of air by $n(r) = \varrho\nu(r)$ where ϱ is the air density.) At a later time t the droplets will have grown by condensation and the distribution will have changed as indicated schematically in Fig. 7.5. To obtain an expression for the variation of $\nu(r, t)$ with time, we consider the rate at which the number of droplets in radius interval δr is changing. We can think of the droplets as flowing through r space as they grow. Let $I(r, t)$ denote the droplet current or number of droplets per unit time per unit mass of air passing the point r in radius space. Then the rate of change of the number of droplets in δr is given by

$$\frac{\partial}{\partial t}(\nu \delta r) = -\frac{\partial I}{\partial r}\delta r. \qquad (7.30)$$

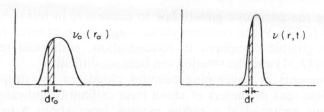

FIG. 7.5. Evolution of droplet spectrum by condensation (schematic).

The droplet current is accounted for entirely by condensation in the conditions assumed here, so that

$$I(r,\ t) = v\,\frac{dr}{dt}.$$

Substituting this equation in (7.30) and dividing both sides by δr gives

$$\frac{\partial v}{\partial t} = -\frac{\partial}{\partial r}\left(v\,\frac{dr}{dt}\right) \qquad (7.31)$$

as the equation describing the change in the droplet size distribution. It is essentially the continuity equation for droplets. For a steady supersaturation, the growth rate is given by the parabolic law $dr/dt = \xi/r$, where $\xi = (S - 1)/(F_k + F_d)$. It may then be confirmed by substitution that the solution of (7.31) is

$$v(r,\ t) = \frac{r}{\sqrt{r^2 - 2\xi t}}\,v_0(\sqrt{r^2 - 2\xi t}). \qquad (7.32)$$

This result expresses the distribution at any time t in terms of the initial distribution v_0 and the growth parameter ξ.

Equation 7.31 is a special case of what is called by Friedlander (1977) the general dynamic equation, which describes the evolution of a particle size distribution by many processes, of which condensation is only one. The solution (7.32) can actually be obtained without recourse to the differential equation (7.31) that defines the process. Because growth by condensation is the only process considered, no drops are created by nucleation or lost by precipitation, coagulation, or diffusion out of the cloud. In Fig. 7.5, the number of droplets in the interval dr at time t is therefore the same as the number in the interval dr_0 at the initial time. That is,

$$v(r,\ t)dr = v_0(r_0)dr_0.$$

Thus

$$v(r,\ t) = v_0(r_0)(dr_0/dr).$$

Employing the parabolic growth law to express r_0 in terms of r leads directly to (7.32). When sedimentation, coagulation, or other processes are not negligible compared to condensation, additional terms are required in (7.31) and the solution can become difficult.

Measurements in developing cumulus clouds of the droplet sizes within a few tens of meters of cloud base ordinarily indicate narrow distributions centered at a radius ranging from about 5 to 10 μm, increasing with distance from cloud base, which are reasonably consistent with the theory of growth by condensation. Higher in the cloud, and in later stages of development, droplet spectra are broader, and may extend to larger drop sizes, than predicted by condensation growth in an ascending, unmixed cloud parcel. Effects in addition to condensation account for spectral broadening, and will be discussed in the next chapter.

The theory of growth by condensation–diffusion, as presented up to now, rests on several approximations and assumptions that are not always satisfied. In the remainder of this chapter we will assess the importance of some of these approximations, correcting the elementary theory where possible to allow for effects that were neglected earlier.

Some corrections to the diffusional growth theory

(a) Kinetic effects

The mechanisms of heat, mass, and momentum transfer between a drop and its surroundings depend on the Knudsen number, defined as the ratio of the molecular mean free path of the gas to the radius of the drop. The mean free path in air, l, varies approximately inversely with the density and is about $0.06\,\mu$m for normal conditions at sea level. When $l/r \ll 1$, as for large drops, the fields of vapor and temperature may be regarded as continua. The equations derived earlier for the transfer of heat and vapor are based on the Maxwell continuum approximation and hence are valid for drops that are much larger than the mean free path.

In the other extreme, when $l/r \gg 1$, the exchange of heat and mass can be calculated from molecular collision theory. This situation, appropriate for particles in the air much smaller than $0.06\,\mu$m, is known as free molecular flow. Small, newly generated cloud droplets have sizes typically between 0.1 and 1 μm radius, for which neither the free molecular nor the continuum approximation is valid. Unfortunately, there is no complete theory for heat and mass transfer for Knudsen numbers near unity. Approximations are derived by interpolating between the free molecular theory and the continuum theory, but uncertainties remain in the quantitative results that follow from the approximations.

Fukuta and Walter (1970) developed a method that is now widely used to allow for kinetic effects in droplet growth calculations. Their approach requires two new parameters that characterize the molecular transfer of heat and vapor: the accommodation coefficient α and the condensation coefficient β. The accommodation coefficient is defined by

$$\alpha = \frac{T_2' - T_1}{T_2 - T_1},$$

where T_2' is the temperature of the vapor molecules leaving the surface of the liquid, T_2 is the temperature of the liquid, and T_1 is the temperature of the vapor. α may be thought of as the fraction of the molecules bouncing off the surface of the drop that have acquired the temperature of the drop. The condensation coefficient is defined as the fraction of the molecules hitting the liquid surface that condense. α and β are, in general, not equal and must be determined experimentally. Within a distance of approximately l from the surface, it is assumed that the mass flow of vapor molecules per unit area and time is controlled by kinetic effects and given by

$$(dm/dt)_k = \frac{\beta(e - e_s)}{(2\pi R_v T)^{1/2}}, \qquad (7.33)$$

where e is the ambient vapor pressure, e_s is the saturation vapor pressure at the surface of the drop, and T is the ambient temperature. Fukuta and Walter argue that the flux given by the kinetic equation must be the same as that given by Maxwell's theory, for a suitably adjusted value of the diffusion coefficient. Equating (7.33) and (7.7), they show that

$$dm/dt = 4\pi r g(\beta) D(\varrho_v - \varrho_{vr}), \qquad (7.34)$$

where $g(\beta)$ is a normalization factor given by

$$g(\beta) = \frac{r}{r + l_\beta}, \qquad (7.35)$$

with l_β a length scale defined by

$$l_\beta = \frac{D}{\beta} \left(\frac{2\pi}{R_v T} \right)^{1/2} \qquad (7.36)$$

A similar analysis of the transfer of heat leads to

$$dQ/dt = 4\pi r f(\alpha) K(T_r - T), \qquad (7.37)$$

where the normalization factor is

$$f(\alpha) = \frac{r}{r + l_\alpha}, \qquad (7.38)$$

with

$$l_a = \left(\frac{K}{\alpha p}\right) \frac{(2\pi R' T)^{1/2}}{(c_v + R'/2)} \tag{7.39}$$

where R' is the gas constant of air, c_v its specific heat at constant volume, and p the pressure.

When the kinetic effects are included, the growth equation becomes

$$r\frac{dr}{dt} = \frac{S - 1}{\left[\left(\dfrac{L}{R_v T} - 1\right)\dfrac{L\varrho_L}{KTf(\alpha)} + \dfrac{\varrho_L R_v T}{De_s(T)g(\beta)}\right]}. \tag{7.40}$$

For small drops, $f(\alpha)$ and $g(\beta)$ are less than unity and the kinetic effects are a barrier to growth. As r increases, these factors approach unity and (7.40) reduces to the continuum solution, (7.17).

The reduction in growth rate depends on the values of the coefficients α and β. Houghton (1985) explains that the values are poorly known, but that α is thought to be close to unity whereas the estimates of β range from 0.02 to 0.04. Figure 7.6 compares the growth of drops by condensation with and without the kinetic corrections. Two examples are given, for drops initially with radii of 1 μm and 5 μm. For all calculations the conditions are $\alpha = 1$, $\beta = 0.04$, $p = 100$ kPa, $T = 283$ K, and $s = 0.5\%$. The solution and curvature effects in the growth equation are neglected. The results show that inclusion of the kinetic corrections reduces the size reached by a drop at any time after it starts to grow, but that the rate of growth after sufficient elapsed time becomes the same as in the con-

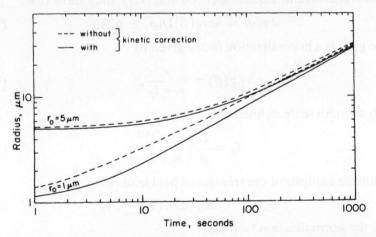

FIG. 7.6. Comparison of condensation growth with and without kinetic corrections. Two examples are shown, for initial radii of 1 μm and 5 μm.

tinuum approximation. Because the growth of the smaller drop is retarded more than that of the larger one, the kinetic corrections account for a relatively broader spectrum at subsequent times than the continuum approximation.

Fitzgerald (1972) included the kinetic corrections in a comparison of measured and calculated droplet spectra in developing cumulus clouds. The data consisted of horizontally averaged updraft velocity measured by airplane several hundred feet below cloud base, the size distribution of NaCl nuclei inferred from a measured activity spectrum of condensation nuclei, and the measured temperature and pressure at cloud base. Using the growth equation in the form (7.40), Fitzgerald computed the drop spectrum at a height of a few hundred meters above cloud base and compared results with observed spectra. He made the comparisons for five continental and two maritime clouds. Both the observed and computed spectra were narrowly peaked at a dominant size. Close agreement was found, as indicated by drop concentrations, the mean drop size, and the standard deviation of drop size. The average values of the dispersion coefficient (the ratio of standard deviation to mean drop diameter) were 0.17 for the observed distributions and 0.12 for the computed distributions. An example of the results for one of the continental clouds is shown in Fig. 7.7.

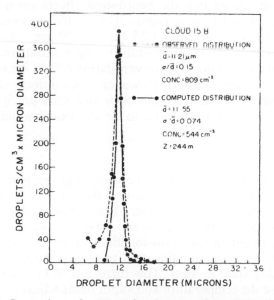

FIG. 7.7. Comparison of computed and measured cloud droplet spectrum at height of 244 m above cloud base. (From Fitzgerald, 1972.)

(b) Ventilation effects

We have assumed that the vapor field surrounding each drop is spherically symmetrical. This is appropriate for a droplet at rest, but when the drop is large enough to fall through the surrounding air with a significant speed, ambient air is continually replenished in the vicinity of the drop, and the vapor field becomes distorted. The rates of heat and mass transfer increase, and are greatest on the upstream side of the drop. These effects are incorporated into the theory of diffusional growth by multiplying the right-hand sides of (7.7) and (7.8) by appropriate ventilation coefficients that are experimentally determined and based on aerodynamic theory. The coefficients equal unity for a droplet at rest and increase with increasing fall speed. The coefficients need not be the same for heat and for vapor transfer, though they are often assumed to be so as a convenient approximation. From results given by Pruppacher and Klett (1978), the ventilation coefficient for vapor transfer may be written

$$f = \begin{cases} 1.00 + 0.09Re, & 0 \leq Re \leq 2.5 \\ 0.78 + 0.28Re^{1/2}, & Re > 2.5 \end{cases}$$

where Re is the Reynolds number of the flow around the drop, defined by $Re = 2\varrho r u/\mu$, in which r is the drop radius and u its fall speed, and ϱ and μ denote the density and dynamic viscosity of the air. Evaluating f for typical conditions shows that the ventilation effects are negligible for droplets smaller than 10 μm in radius. The coefficient equals approximately 1.06 for $r = 20$ μm and 1.25 for $r = 40$ μm. The growth by condensation of drops this large is usually negligible compared to growth by collision and coalescence. Hence ventilation effects are unimportant in the growth of cloud drops. However, such effects can be very significant in the evaporation of raindrops and other precipitation particles that have high velocities relative to the air through which they fall.

(c) Nonstationary growth

The fields of vapor and temperature about a growing or evaporating drop cannot be exactly steady because the surface of the drop is expanding or contracting. Moreover, the vapor content of the air surrounding the drop may change as the drop moves around, requiring continual adjustment of the diffusion field in its immediate vicinity. Nevertheless it is a great simplification to regard the flow of heat and vapor as a steady state process. This turns out to be a good approximation, because the temperature and vapor fields adjust very quickly to the presence of a drop and assume the steady state configuration.

Borovikov *et al.* (1961) gave a nonsteady solution of the diffusion equation (7.1) corresponding to the situation in which a drop is inserted into an initially uniform vapor field. At time zero the concentration of vapor molecules $n(R)$ is constant at the undisturbed value n_0, for all distances $R > r$, the drop radius. At all times the vapor concentration at the surface of the drop is the equilibrium value, n_r. The solution satisfying these conditions is

$$n(R) = n_\infty - \frac{r}{R} (n_\infty - n_r)\left[1 - \text{erf}\left(\frac{R - r}{2\sqrt{Dt}} \right) \right],$$

where erf (x) denotes the error function. Comparing the growth rate for this general solution with that derived from (7.4) for steady conditions, they found that after only 10 μs, for droplets of typical size, the rates of growth are essentially the same. Molecular diffusion thus quickly establishes a field around the drop that corresponds to the steady state solution. Consequently the transient response of the field may be neglected in calculations of drop growth and evaporation.

(d) Unsteady updraft

In an attempt to explain the eventual broadening of droplet spectra with increasing distance above cloud base, several researchers (e.g., Warner, 1969b) have calculated the growth of a family of drops by condensation in a time-varying updraft. They find that the resulting drop spectra often are not significantly broader than those produced by a steady updraft. Some updraft structures, however, especially those that include a general acceleration, are capable of broadening the spectrum by continual activation of fresh nuclei, leading to a persistent bimodal drop-size distribution. In his calculations, Warner (1969b) found that the condensation coefficient β could not be much larger than 0.05 to account for the almost invariable observations of significant numbers of droplets with $r < 5 \mu$m at heights of 100 m or so above cloud base. But for droplets to grow beyond this size in reasonable times the coefficient cannot be much less than 0.03.

Manton (1979) explained that calculations of droplet evolution in unmixed, ascending parcels will inevitably give results that are relatively insensitive to the assumed updraft structure because of the high correlation between supersaturation and vertical velocity. Thus, in a weak updraft, the droplets have a long time to grow before reaching a given altitude, but the supersaturation is low. The shorter time to reach the same altitude in a strong updraft is approximately compensated by the higher supersaturation. Except for the creation of newly activated droplets in regions of upward acceleration, unsteady updrafts in unmixed parcels have little effect on the droplet evolution.

(e) Statistical effects

Calculations of droplet spectrum evolution in unmixed parcels of air, even with an unsteady updraft to provide a time-varying supersaturation, do not embody the variability of natural clouds. Mixing processes and the gravitational settling of drops (sedimentation) prevent a population from staying together indefinitely. Therefore different drops will ultimately have different histories. It seems very plausible that some, by chance, will grow faster than others by experiencing higher than average supersaturations or longer residence times during their development. The droplets in a cloud sample taken some distance above cloud base or some time after the initial condensation occurs are thus likely to have a distribution broader than those growing in an unmixed parcel.

A group of Soviet cloud physicists, starting with work as reported by Mazin (1968) and recently typified by that of Smirnov and Nadeykina (1984), have developed a theory of droplet growth by condensation that includes the effects of turbulence-induced fluctuations in supersaturation and diffusive mixing. Called stochastic condensation, the theory accounts for some broadening of the size distribution of growing drops, depending on the strength of the turbulent mixing. The theory has not been universally embraced, probably because the amount of spectral broadening it predicts can be achieved by evidently simpler kinds of large scale mixing, as will be discussed in the next chapter. Only recently, however, Cooper *et al.* (1986) have reported on airplane measurements in Hawaiian orographic clouds that indicate high variability in the supersaturation, with values as large as 2% in 2 to 3% of the cloud volume. They suggest that such high variability might make it possible for a few droplets to grow much more rapidly than the average, though whether the observations support the Soviet theory of stochastic condensation or alternative theories is not clear.

It has been speculated (Srivastava, 1969) that droplets may grow at different rates even in a homogeneous vapor field because the droplet concentration in a well mixed cloud is a random variable with a Poisson probability distribution. Thus, if the average concentration is, say, 500 per cm^3, the local concentration will vary considerably about this value. On average, each drop occupies a volume of $1/500 = 0.002 \, cm^3$, but some fraction of the drops occupy or control a volume less than this amount. If the local concentration were sufficiently high, then the volume available to each drop could conceivably be so small that the vapor field would not achieve its theoretical steady state profile outwards from the drop surface. Neighboring drops would distort the field, retarding drop growth. This reduction would have to be compensated by a faster growth rate of the drops in areas of low concentration. Otherwise, the supersaturation would increase.

In spite of the reality of these fluctuations, there is reason to expect that the effect on droplet growth may be small. From (7.4) for the steady state vapor profile, it is clear that the influence of a drop on the vapor field becomes small at outward distances exceeding ten radii or so. Yet even in regions of high droplet concentration due to chance fluctuations, the droplet separation is nearly everywhere greater than this amount. Kabanov *et al.* (1971) gave more detailed arguments to conclude that the natural fluctuations in droplet concentration should have a negligible influence on the evolving drop spectrum.

The understanding of drop spectral broadening after the initial stage of condensation is a challenging problem that has occupied many researchers over the past two decades. It will be discussed further in the next chapter in the context of precipitation development.

Problems

7.1. Prove that, in general, the difference between the masses of two droplets growing in the same environment according to the approximation (7.19) increases with time.

7.2. A droplet grows by condensation from a radius of 2 μm to 20 μm in 10 min at a temperature of 0°C and a pressure of 70 kPa. Estimate the ambient supersaturation, neglecting the solution and curvature terms in the growth equation.

7.3. Show that the pseudoadiabatic lapse rate (eq. 3.16) can be derived from equations (7.22)–(7.24).

7.4. A sample of moist air is cooled isobarically. A cloud forms and the cooling continues. In a form similar to (7.22), the rate of change of the saturation ratio may be written

$$\frac{dS}{dt} = q_1 \frac{dT}{dt} - q_2 \frac{d\chi}{dt},$$

where dT/dt and $d\chi/dt$ are the rates of change of temperature and condensed water. Derive expressions for the thermodynamic factors q_1 and q_2. Evaluate these expressions for $p = 80$ kPa and $T = 280$ K.

7.5. To analyze the approximate behavior of the supersaturation in a cloud of growing droplets, suppose the droplets are all the same size, growing by condensation according to (7.19). Let ν_0 denote their concentration per unit mass of air, and regard this quantity as a constant under the assumption that no new drops are created and that no existing drops are lost. Show that these assumptions, taken in connection with (7.22), lead to (7.29). Assuming further that the temperature, pressure, and droplet size are slowly varying compared to the supersaturation s, find the solution of this equation that satisfies the initial condition $s = s_0$ at $t = 0$. Show that the supersaturation tends to $s_\infty = \omega/\eta$ as the time increases, and that the relaxation time of the supersaturation is η^{-1}.

7.6. The two droplets in problem 6.3 grow in a steady environment with $p = 80$ kPa, $T = 280$ K, and $s = 0.8\%$. Calculate their sizes as a function of time starting at $r = r^*$ for each droplet and continuing for 20 min. An accurate solution to this problem requires careful numerical integration of the diffusional growth equation, (7.18). Compare these results with an analytical solution of the growth of a droplet in the same environment starting at size $r = 0$, neglecting the solution and curvature terms in the growth equation.

7.7. In a developing cumulus cloud the droplet spectrum has a Gaussian shape, centered at a mean radius of 4 μm with a standard deviation of 0.6 μm. Given that the cloud water content is 0.2 g/m^3, estimate the supersaturation in an updraft of 8 m/s. Assume a temperature of 0°C and a pressure of 80 kPa.

8

Initiation of Rain in Nonfreezing Clouds

Most of the world's precipitation falls to the ground as rain, much of which is produced by clouds whose tops do not extend to temperatures colder than 0°C. The mechanism responsible for precipitation in these "warm" clouds is coalescence among cloud droplets. The dominant precipitation-forming process in the tropics, coalescence is also effective in some midlatitude cumulus clouds whose tops may extend to subfreezing temperatures.

A central task of precipitation physics is to explain how raindrops can be created by condensation and coalescence in times as short as 20 min. This is often quoted as the interval observed between the initial development of a cumulus cloud and the first appearance of rain. During this time a cloud population, consisting of the order of 100 droplets per cm^3 averaging 10 μm in radius, evolves into a raindrop population of 1000 drops per m^3 with typical diameters of 1 mm. Although there are still some uncertainties about details, it is generally agreed that collisions and coalescence of droplets account for nearly all of this 50-fold increase in radius, but that these events do not occur in significant numbers until some of the droplets grow to a radius of about 20 μm. Smaller droplets have small collision cross sections and slow settling speeds and hence have little chance of colliding with one another. Coalescence can only become significant after the droplet spectrum evolves to include a spread of sizes and fall speeds, with some droplets reaching 20 μm or so. The rate of collisions that a droplet of this size experiences increases rapidly with size, in proportion to at least the fourth power of the radius, so that coalescence proceeds at an accelerating pace once it begins. By the time some droplets reach a radius of 30 μm in a developing cloud, coalescence is likely to be the dominant growth process. A typical raindrop of 1 mm diameter may be the result of the order of 10^5 collisions. Before these can begin, some other process must account for the production of a few droplets (perhaps only one in 10^5 droplets, or one per liter of cloud volume) as large as 20 μm in radius.

Setting the stage for coalescence

From the approximate equation (7.19), it may be determined that a cloud droplet can grow by condensation to a radius of 20 μm in 10 min under conditions of constant ambient saturation ratio if the super-saturation is about 0.5%. For typical droplet concentrations, this super-saturation would require a sustained updraft of 5 m/s or greater. Such special conditions would only be approximated in developing cumulus clouds of considerable vertical extent, but in these clouds it appears plausible that condensation in an adiabatically ascending cloudy parcel can produce the droplets needed to initiate coalescence in realistic times.

Condensation alone, however, does not account for the broadening of the droplet spectrum that seems to be concomitant with the growth that occurs before coalescence begins. A broadening toward larger sizes greatly increases the chances of coalescence once a size approaching 20 μm is reached. A broadening to smaller sizes contributes little to coalescence because the effective collision cross sections remain very small.

For a time it was thought that giant sea salt nuclei, or possibly giant insoluble particles, might account for the one particle in 10^5 needed to initiate coalescence. There is no evidence that such outliers from the regular cloud population are always present, and even where they are known to exist it now appears that they may not be essential for the formation of rain (Woodcock *et al.*, 1971). Other explanations must therefore be sought for the onset of coalescence.

A growing body of observations and theoretical calculations point to mixing between the cloud and its environment as the most likely explanation for spectral broadening prior to coalescence. Two different kinds of mixing may be envisioned when subsaturated, cloud-free air is entrained into a cloud. In the first, it is assumed that mixing occurs quickly and completely so that all the droplets at a given level in the cloud are exposed to identical conditions of subsaturation. As a group, they evaporate until saturation (the equilibrium condition) is reached. The effects of this process, often called homogeneous mixing, following Baker and Latham (1979), are to reduce all the droplet sizes by evaporation, to reduce the concentration by dilution in proportion to the amount of outside air introduced, and possibly to introduce newly activated droplets, depending on the treatment of nucleation. Homogeneous mixing implicitly assumes that the time required for the cloudy and the entrained air to mix is short compared to the time for the droplets to evaporate and re-establish vapor equilibrium. This process can explain a broadening of the droplet spectrum to smaller sizes, but cannot account for the creation of larger droplets than in an unmixed, adiabatic parcel.

The alternative view of the mixing process supposes that the time scale of droplet evaporation is short compared to that of turbulent mixing.

Evaporation proceeds rapidly in the cloud regions first exposed to the entrained air, creating volumes of air that are droplet-free but essentially saturated. These volumes then mix with the previously unaffected cloud, reducing the concentration of droplets by dilution without changing their size. This process is sometimes called inhomogeneous mixing, although it should be noted that the initial evaporation step is the same as for homogeneous mixing. The difference is that in this alternative view the evaporation step is applied to relatively small volumes. A cloud may be regarded as consisting of many such volumes with different sizes and mixing histories. It is inhomogeneous on scales larger than these volumes. The important distinction from this point of view is that the entrainment of dry air into a cloud need not lead to a general reduction in droplet size. The cloud water concentration will be reduced from its unmixed value, but depending on the details of the mixing process the drop sizes in some parts of the cloud may not be affected.

Many researchers have recently developed models of cloud droplet evolution including entrainment and turbulent mixing. These have succeeded in explaining certain significant features of observations: the variability of spectral shapes; the variability of water concentration and its reduction below the adiabatic value; the broadening of spectra towards smaller sizes with increasing time and altitude. Some of the key papers are those of Baker and Latham (1979), Telford and Chai (1980), and Jonas and Mason (1982). Paluch and Knight (1984) may be consulted for a review of this work.

Telford and Chai (1980) argued further that the entrainment of dry air at cloud top can explain the production of large droplets. The entrained air mixes with the cloud, evaporating just enough droplets to bring the mixture to saturation. This chilled volume of saturated air, which may contain droplets that were too large to evaporate completely, descends through the cloud as a distinct entity—a cold, saturated stream—maintaining saturation during descent by evaporation of droplets mixed in from the adjacent cloud. This process proceeds down from the cloud top in repeated cycles, progressively diluting the cloud from the top downwards. The cloud thus consists of many distinct entities or blobs, some moving upwards with droplets growing by condensation, others downwards at saturation but with few droplets. The effect of mixing between the upwards and downwards moving entities is to reduce the concentration of droplets in the ascending air. The supersaturation created by the updraft is then distributed over fewer droplets, permitting them to grow to larger sizes than they would otherwise. There is still no consensus on details of the mixing mechanism, as may be seen from recent work (Telford *et al.*, 1984; Paluch and Knight, 1986; Telford, 1987). It seems, however, that the new understanding of the significance of cloud-top entrainment may eventually explain many of the observed microphysical characteristics of clouds.

Droplet growth by collision and coalescence

Collisions may occur through differential response of the droplets to gravitational, electrical, or aerodynamic forces. Gravitational effects predominate in clouds: large droplets fall faster than smaller ones, overtaking and capturing a fraction of those lying in their paths. The electrical and turbulent fields required to produce a comparable number of collisions are much stronger than those thought usually to exist, although the intense electric fields in thunderstorms may create significant local effects. As a drop falls it will collide with only a fraction of the droplets in its path because some will be swept aside in the airstream around the drop. The ratio of the actual number of collisions to the number for complete geometric sweep-out is called the collision efficiency, and depends on the size of the collector drop and the sizes of the collected droplets.

Collision does not guarantee coalescence. When a pair of drops collide several types of interaction are possible: (1) they may bounce apart; (2) they may coalesce and remain permanently united; (3) they may coalesce temporarily and separate, apparently retaining their initial identities; (4) they may coalesce temporarily and then break into a number of small drops. The type of interaction depends upon the drop sizes and collision trajectories, and is also influenced by the existing electrical forces and other factors. For sizes smaller than 100 μm in radius, the important interactions are (1) and (2) in the preceding list. The ratio of the number of coalescences to the number of collisions is called the coalescence efficiency. The growth of a drop by the collision-coalescence process is governed by the *collection efficiency*, which is the product of collision efficiency and coalescence efficiency. Laboratory studies of small colliding droplets indicate that the coalescence efficiency is close to unity if the droplets are charged or an electrical field is present. Because weak fields and charges exist in natural clouds, theoretical studies of droplet growth by collision-coalescence usually make the assumption that the collection efficiency equals the collision efficiency. The problem of explaining the development of rain then reduces to one of determining collision rates among a population of droplets.

(a) Droplet terminal fall speed

The drag force exerted on a sphere of radius r by a viscous fluid is given by

$$F_R = \frac{\pi}{2} r^2 u^2 \varrho C_D, \qquad (8.1)$$

where u is the velocity of the sphere relative to the fluid, ϱ is the fluid density, and C_D is the drag coefficient characterizing the flow. In terms of the Reynolds number $Re = 2\varrho ur/\mu$, with μ the dynamic viscosity of the fluid, (8.1) may be written in the form

$$F_R = 6\pi\mu ru(C_D Re/24).\qquad(8.2)$$

The gravitational force on the sphere is given by

$$F_G = \tfrac{4}{3}\pi r^3 g(\varrho_L - \varrho),$$

where ϱ_L is the density of the sphere. For a water drop falling through air $\varrho_L \gg \varrho$ and

$$F_G = \tfrac{4}{3}\pi r^3 g\varrho_L\qquad(8.3)$$

to good approximation. When $F_G = F_R$ the drop falls relative to the air at its terminal fall speed. For this equilibrium situation

$$u^2 = \frac{8}{3}\frac{rg\varrho_L}{\varrho C_D}$$

or

$$u = \frac{2}{9}\frac{r^2 g\varrho_L}{(C_D Re/24)\mu}.\qquad(8.4)$$

For very small Reynolds numbers the Stokes solution* to the flow field around a sphere shows that $(C_D Re/24) = 1$. In this case, (8.4) reduces to

$$u = \frac{2}{9}\frac{r^2 g\varrho_L}{\mu} = k_1 r^2,\qquad(8.5)$$

with $k_1 \approx 1.19 \times 10^6 \text{ cm}^{-1}\text{s}^{-1}$. This quadratic dependence of fall speed on size is called Stokes' Law and applies to cloud droplets up to about $30\,\mu\text{m}$ radius.

Experiments with spheres indicate that for sufficiently high Reynolds numbers C_D becomes independent of Re and has a value of about 0.45. Using this value in (8.4) leads to

$$u = k_2 r^{1/2},\qquad(8.6)$$

where

$$k_2 = 2.2 \times 10^3\left(\frac{\varrho_0}{\varrho}\right)^{1/2} \text{cm}^{1/2}\text{ s}^{-1}.\qquad(8.7)$$

In (8.7) ϱ is the air density and ϱ_0 is a reference density of 1.20 kg/m^3, corresponding to dry air at 101.3 kPa and 20°C. Raindrops have high Reynolds numbers but are not perfectly spherical. Consequently, though often a useful approximation, (8.6) describes the fall speed of raindrops reasonably well only over a limited range of size.

---

* See, for example, Lamb (1945), pp. 598–599.

TABLE 8.1. *Terminal Fall Speed as a Function of Drop Size (equivalent spherical diameter) (From Gunn and Kinzer, 1949)*

Diam. (mm)	Fall speed (m/s)	Diam. (mm)	Fall speed (m/s)
0.1	0.27	2.6	7.57
0.2	0.72	2.8	7.82
0.3	1.17	3.0	8.06
0.4	1.62	3.2	8.26
0.5	2.06	3.4	8.44
0.6	2.47	3.6	8.60
0.7	2.87	3.8	8.72
0.8	3.27	4.0	8.83
0.9	3.67	4.2	8.92
1.0	4.03	4.4	8.98
1.2	4.64	4.6	9.03
1.4	5.17	4.8	9.07
1.6	5.65	5.0	9.09
1.8	6.09	5.2	9.12
2.0	6.49	5.4	9.14
2.2	6.90	5.6	9.16
2.4	7.27	5.8	9.17

Accurate observational data on raindrop fall speed are those given by Gunn and Kinzer (1949), reproduced here in Table 8.1. These data were obtained at sea-level conditions, 101.3 kPa and 20°C. Owing to the reduced air density, a droplet of given size will tend to fall faster aloft than at sea level, approximately in accordance with the square-root law of (8.7). Beard (1976) developed empirical formulas which fit the data accurately and which can also be corrected for temperature and pressure. At the surface, raindrops which have attained the largest possible size before breakup fall no faster than about 9 m/s. Under typical conditions at 500 mb the upper limit is about 13 m/s. Beard's formulas apply to three separate ranges of diameter and are rather complicated.

It can be determined from the data that (8.6) provides a reasonable approximation to the fall speed in the radius interval $0.6 \, \text{mm} < r < 2 \, \text{mm}$, but with $k_2 \approx 2.01 \times 10^3 \, \text{cm}^{1/2} \, \text{s}^{-1}$. In the intermediate size range, between the Stokes' Law region and the square-root law, an approximate formula for fall speed is

$$u = k_3 r, \qquad 40 \, \mu\text{m} < r < 0.6 \, \text{mm} \tag{8.8}$$

with $k_3 = 8 \times 10^3 \, \text{s}^{-1}$.

(b) Collision efficiency

A drop of radius R is pictured in Fig. 8.1 overtaking a droplet of radius r. If the droplet had zero inertia it would be swept aside by the stream flow around the larger drop and a collision would not occur. Whether a collision does in fact occur depends on the relative importance of the

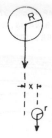

FIG. 8.1. Collision geometry.

inertial and aerodynamic forces, and the separation x between drop centers, called the impact parameter. For given values of r and R there is a critical value x_0 of the impact parameter within which a collision is certain to occur and outside of which the droplet will be deflected out of the path of the drop. Results are presented in the form of collision efficiencies, defined by

$$E(R, r) = \frac{x_0^2}{(R + r)^2} .$$ (8.9)

So defined, the collision efficiency is equal to the fraction of those droplets with radius r in the path swept out by the collector drop that actually collide with it. Alternatively, E may be interpreted as the probability that a collision will occur with a droplet located at random in the swept volume.

Figure 8.2 presents collision efficiencies for small collector drops as obtained from three sets of theoretical calculations. The efficiencies are plotted against the ratio r/R of drop radii for different sizes of the collector drop. Manton (1974) has shown by dimensional arguments that the only important parameters affecting the collision of water drops in air are r/R and gR^3/v^2, where g is gravity and v is the kinematic viscosity.

The calculations by Schlamp *et al.* (1976) and Lin and Lee (1975) are based on the superposition method, which assumes that each drop moves in a flow field generated by the other drop moving in isolation. This procedure was first suggested by Irving Langmuir, a versatile industrial researcher who won the Nobel Prize in Chemistry in 1932 for work on surface kinetics, and who in later years turned his interest to problems in atmospheric physics. It is the only accurate method available when the Reynolds number exceeds unity. In the calculation of Klett and Davis (1973) the complete equation governing the flow around the two drops is used but the inertial force is only approximated.

For any size of collector drop, the collision efficiency is small for small values of r/R. The collected droplets are then small, have little inertia, and are easily deflected by the flow around the collector drop. The inertia

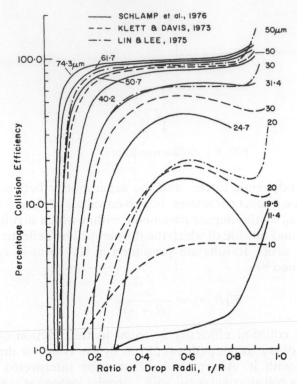

FIG. 8.2. Computed collision efficiencies for pairs of drops as a function of
the ratio of their radii. Curves are labeled according to the radius R of the
larger drop.

of the droplets increases with r/R, accounting for an increase in collision
efficiency up to a radius ratio of about 0.6. Two counteracting effects
come into play as r/R increases beyond this value. Because the difference
in the size of the drops is getting smaller, the relative velocity between the
drops is reduced, prolonging the time of interaction. The flow fields
interact strongly, and the time can be sufficient for the droplet to be
deflected around the drop without collision. On the other hand, there is
a possibility for a trailing droplet to be attracted into the wake of a drop
falling close by at nearly the same speed. This effect can lead to "wake
capture" and to collision efficiencies that exceed unity for values of
$r/R \approx 1$. The theoretical prediction that E can exceed 1.0 for radius
ratios near unity is supported by laboratory experiments. However,
owing to the large number of integration steps required in the computation of collisions between drops of nearly equal size, the calculations of
collision efficiency by wake capture are of doubtful accuracy.

These results for relatively small collector drops, as well as the best

TABLE 8.2. *Collision Efficiency* E *for Drops of Radius* R *Colliding with Droplets of Radius* r. *(Data for* R < 50 μm *adapted from Klett and Davis, 1973; for 50 μm* ≤ R ≤ *500 μm from Beard and Ochs, 1984; for* R > *500 μm from Mason, 1971)*

R (μm)	2	3	4	6	8	10	15	20	25
				$r(\mu m)$					
10	0.017	0.027	0.037	0.052	0.052				
20	*	0.016	0.027	0.060	0.12	0.17	0.17		
30	*	*	0.020	0.13	0.28	0.37	0.54	0.55	0.47
40	*	*	0.020	0.23	0.40	0.55	0.70	0.75	0.75
50	—	—	0.030	0.30	0.40	0.58	0.73	0.75	0.79
60	—	0.010	0.13	0.38	0.57	0.68	0.80	0.86	0.91
80	—	0.085	0.23	0.52	0.68	0.76	0.86	0.92	0.95
100	—	0.14	0.32	0.60	0.73	0.81	0.90	0.94	0.96
150	0.025	0.25	0.43	0.66	0.78	0.83	0.92	0.95	0.96
200	0.039	0.30	0.46	0.69	0.81	0.87	0.93	0.95	0.96
300	0.095	0.33	0.51	0.72	0.82	0.87	0.93	0.96	0.97
400	0.098	0.36	0.51	0.73	0.83	0.88	0.93	0.96	0.97
500	0.10	0.36	0.52	0.74	0.83	0.88	0.93	0.96	0.97
600	0.17	0.40	0.54	0.72	0.83	0.88	0.94	0.98	†
1000	0.15	0.37	0.52	0.74	0.82	0.88	0.94	0.98	†
1400	0.11	0.34	0.49	0.71	0.83	0.88	0.94	0.95	†
1800	0.08	0.29	0.45	0.68	0.80	0.86	0.96	0.94	†
2400	0.04	0.22	0.39	0.62	0.75	0.83	0.92	0.96	†
3000	0.02	0.16	0.33	0.55	0.71	0.81	0.90	0.94	†

—Collision efficiency less than 0.01.
*Value cannot be determined accurately from available data.
† Value close to one.

available data for larger collector drops, are summarized in Table 8.2. The entries in this table were used to prepare Fig. 8.3, the field of collision efficiency as a function of R and r. The figure shows that E is generally an increasing function of R and r, but for R greater than about 100 μm E depends largely on r.

Some of the papers on collision efficiency refer to a quantity called the linear collision efficiency, defined by

$$y_c = x_0/R. \qquad (8.10)$$

From (8.9) and (8.10),

$$E = y_c^2/(1 + p)^2, \qquad (8.11)$$

where $p = r/R$. Alternatively the efficiency is sometimes defined by

$$E' = x_0^2/R^2.$$

Since $E' = E(1 + p)^2$, it is clear that E' can take on values greater than unity.

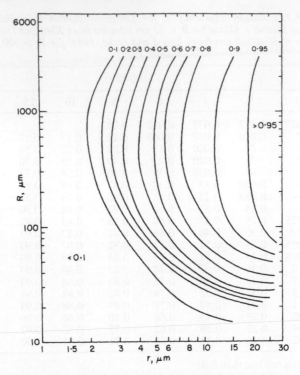

FIG. 8.3. Field of $E(R, r)$ based on data in Table 8.2.

(c) Growth equations

Suppose a drop of radius R is falling at terminal speed through a population of smaller droplets. During unit time it sweeps out droplets of radius r from a volume given by

$$\pi(R + r)^2[u(R) - u(r)]$$

where u denotes terminal fall speed. Thus the average number of droplets with radii between r and $r + dr$ collected in unit time is given by

$$\pi(R + r)^2[u(R) - u(r)]n(r)E(R, r)dr$$

where $E(R, r)$ denotes the *collection* efficiency, which equals the product of collision efficiency and coalescence efficiency. When the drops are all smaller than about 100 μm it is usually assumed that the coalescence efficiency is unity, so that the collection efficiency is identical to the collision efficiency.

The total rate of increase in volume of the collector drop is obtained by integrating over all droplet sizes:

$$\frac{dV}{dt} = \int_0^R \pi(R + r)^2 \frac{4}{3} \pi r^3 E(R, r)n(r)[u(R) - u(r)]dr. \quad (8.12)$$

In terms of drop radius

$$\frac{dR}{dt} = \frac{\pi}{3} \int_0^R \left(\frac{R + r}{R}\right)^2 [u(R) - u(r)]n(r)r^3 E(R, r)dr. \quad (8.13)$$

The change of drop size with altitude may be obtained from

$$\frac{dR}{dz} = \frac{dR}{dt} \frac{dt}{dz} = \frac{dR}{dt} \frac{1}{U - u(R)}. \quad (8.14)$$

(8.13) is general because it allows for the sizes and fall speeds of the collected droplets. If these droplets are much smaller than the collector drop, then an approximation to (8.13) follows by setting $u(r) \approx 0$ and $R + r \approx R$. Thus

$$\frac{dR}{dt} = \frac{\bar{E}M}{4\varrho_L} u(R), \quad (8.15)$$

where $\bar{E}$ is an effective average value of collection efficiency for the droplet population and M is the cloud liquid water content in units of mass per unit volume. From (8.14), the change of radius with altitude is then given approximately by

$$\frac{dR}{dz} = \frac{\bar{E}M}{4\varrho_L} \frac{u(R)}{U - u(R)}. \quad (8.16)$$

If the updraft is negligibly small, then

$$\frac{dR}{dz} = -\frac{\bar{E}M}{4\varrho_L}.$$

These equations describe drop growth as a continuous collection process, regarding the cloud as a continuum. Growth actually occurs by the discrete events of droplet capture. The difference can be significant, as will be explained later.

The Bowen model

The early radar pioneer E. G. Bowen, in an influential paper of 1950, used (8.16) in an assessment of rain development in warm clouds. He assumed that a cloud of uniformly-sized droplets was ascending with a constant updraft and growing by condensation. In addition, he considered that a drop with twice the mass of the others was present as a result of the chance coalescence of two of the droplets. This drop is carried upward with the droplets and allowed to grow by condensation

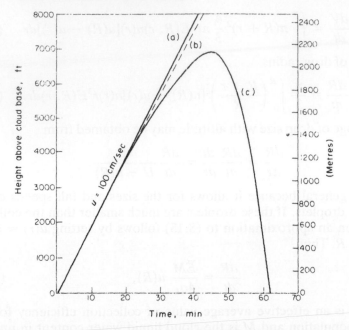

FIG. 8.4. Bowen's calculated trajectories of (a) the air, (b) cloud droplets, initially 10 μm in radius, and (c) drops which have initially twice the mass of the cloud droplets. Updraft speed 1 m/sec, cloud water content $M = 1$ g/m^3. (From Fletcher, 1962.)

and coalescence, using the Langmuir collision efficiencies and appropriate approximations for the terminal fall speed in (8.16). An example of his results is shown in Fig. 8.4.

Growth by coalescence is slow at first and only of the same order of magnitude as that by condensation. However, u and E increase rapidly with drop size and coalescence quickly outpaces condensation. When it has grown to a size that will just be balanced by the updraft, the drop reaches the top of its trajectory. On further growth it begins to descend, growing more on its downward passage, finally emerging as a raindrop from the base of the cloud.

Important parameters in the Bowen model are the updraft velocity and the cloud water content. With increasing updraft velocity, the drop ascends to a higher level before beginning its descent, and emerges from cloud base with a larger size. For a given updraft velocity, drops grow larger but have a lower trajectory as the cloud water content is increased.

To illustrate the effect of the updraft, Figs. 8.5 and 8.6 have been prepared using more recent data on collection efficiency. As in Bowen's calculations, $r = 10\ \mu$m and $M = 1$ g/m^3, but to insure some growth of the collector drop its initial radius was set at 20 μm. Figure 8.5 shows the

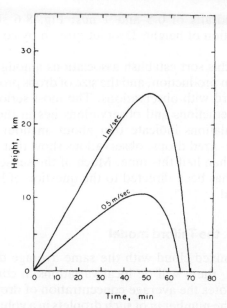

FIG. 8.5. Drop trajectories calculated for the collision efficiencies of Table 8.2 and Fig. 8.3, assuming a coalescence efficiency of unity. Initial drop radius 20 μm. Cloud water content 1 g/m^3; all cloud droplets of 10 μm radius.

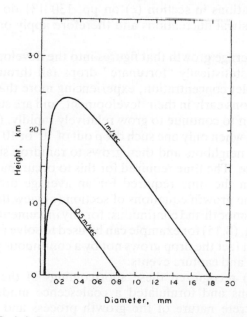

FIG. 8.6. Drop diameters for the trajectories of Fig. 8.5.

trajectories for updrafts of 0.5 and 1 m/s; Fig. 8.6 shows the drop diameter as a function of height. Droplet growth by condensation was neglected.

Calculations of this sort establish associations among updraft, cloud height, time for rain production, and the size of drops produced, that are in qualitative accord with observations. The most serious discrepancy between model predictions and observations lies in the time requirement. While calculations indicate that about an hour is required to produce millimeter-sized drops, observations show that such drops can be formed in less than half this time. Much of the subsequent work on rain development has been directed to the question of how the growth time can be reduced.

Statistical growth: the Telford model

Even in a well-mixed cloud with the same average droplet concentration throughout, there will be local variations in concentration. In particular, if $\bar{n}$ denotes the average concentration of droplets in a given size interval, then the number m of such droplets in a volume $\mathcal{V}$ obeys the Poisson probability law,

$$p(m) = e^{-\bar{n}\mathcal{V}} \frac{(\bar{n}\mathcal{V})^m}{m!}. \tag{8.17}$$

The growth equations in section (c) on pp. 130–131 do not take into account the statistical fluctuations and therefore apply only to average droplet growth.

It is not the average growth that figures into the development of rain, however. Some statistically "fortunate" drops fall through regions of locally high droplet concentration, experiencing more than the average number of collisions early in their development, and are subsequently in a favored position to continue to grow relatively rapidly. (See Fig. 8.7.) Rain is produced when only one such drop out of 10^5 or 10^6 gets an initial head start on its neighbors and then grows to raindrop size by gravitational coalescence. The time required for this to occur may be considerably shorter than the time required for an average droplet to reach raindrop size. The growth equations of section (c) view the coalescence process as being smooth and continuous: for any increment of time Δt, no matter how small, (8.15) for example can be used to solve for the increase in drop radius. In fact the drop grows not by a continuous process but by discrete collision and capture events.

Telford (1955) recognized these shortcomings of the "continuous-growth" equations and formulated a coalescence model taking into account the discrete nature of the growth process and the statistical fluctuations of droplet concentration. He assumed that the collected

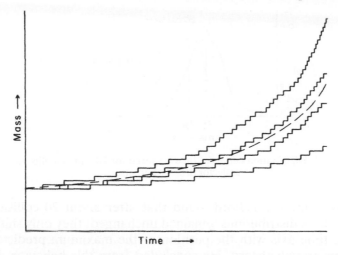

FIG. 8.7. Schematic illustration of droplet growth by discrete captures, including statistical variability. Shown are four realizations of the growth of a drop by collection of smaller, equal-sized droplets. The average growth for many such realizations is indicated by the dashed line.

droplets were all the same size (10 μm radius in the most relevant case considered) and that the collector drops have twice the volume (12.6 μm radius). From the results Telford concluded that the statistical-discrete capture process is crucial in the early stages of rain formation. From the initial bimodal drop-size distribution, reasonable raindrop spectra evolved over periods of a few tens of minutes. Moreover, in one fairly representative case, he found that 100 of the 12.6 μm drops per cubic meter would experience their first 10 coalescences after a time of only about 5 min. Compared to this, the time required for 100 drops/m^3 to undergo 10 collisions under the same conditions but assuming continuous growth would be 33 min.

In his analysis, Telford derived first the probability distribution of the time required for the collector drops to experience a given number of collisions. Examples of such distributions are sketched schematically in Fig. 8.8. Here Mdt represents the proportion of drops that experience n captures ($n = 20, 30$ in this illustration) in the time between t and $t + dt$. The most likely times required for n captures correspond to the peak values of the distributions and are indicated by t_{20} and t_{30} in this example. The most likely time for a given number of collisions is very nearly equal to the time predicted by the continuous-growth equations. The distributions are slightly skewed, however, with the result that the average time for n collisions exceeds somewhat the time for continuous growth to the same size. The drops that grow faster than the average and can account for the relatively rapid development of rain are those in the left-hand tails

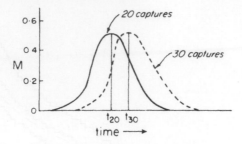

FIG. 8.8. Distribution curves of time required for 20 and 30 collisions of a collector drop (schematic).

of the distributions. Telford found that after about 20 collisions the *shapes* of these distributions remained unchanged; they only shifted out along the time axis with the position of the maximum predictable by continuous-growth theory. He concluded from this behavior that the statistical effects were important only for about the first 20 collisions, by which time the distribution was established and after which the continuous-growth equations could be applied.

Because reliable information on the collision efficiencies of small drops was not available at the time, Telford assumed $E = 1$ in his calculations. This assumption made it possible to solve the equations analytically, but it is now known to be greatly in error for the small drop sizes considered. Using basically the same approach as Telford, Robertson (1974) estimated the importance of statistical effects in coalescence growth, employing accurate data on collision efficiency. Owing to the complex form of $E(R, r)$ (see Fig. 8.3) analytical solutions are not possible and Robertson used a Monte Carlo procedure to simulate the collisions of a collector drop. Like Telford, Robertson found that the time distributions approach a limiting form as the number of collisions increases. Several simulations were carried out, with cloud droplet sizes r ranging between 8 and 14 μm in radius, and collector drops with initial sizes $R(0)$ ranging from 20 to 40 μm. An example of Robertson's results is shown in Fig. 8.9.

As n increases, the standard deviations increase and approach a limiting value. From Fig. 8.9 and results for the other cloud droplet sizes, it was found that the limiting value is achieved after a sufficient number of collisions of the collector drop: approximately 6 for $R(0) = 20 \mu$m, 40 for $R(0) = 30 \mu$m, and 100 for $R(0) = 40 \mu$m. The limiting values of σ decrease with increasing $R(0)$, and are negligibly small for drops greater than 40 μm in radius, implying that the continuous growth equations may be used. For $R(0)$ less than 20 μm, the collision efficiencies are too small to allow development of drizzle drops in times less than several hours, even though the statistical deviations from average drop behavior are large.

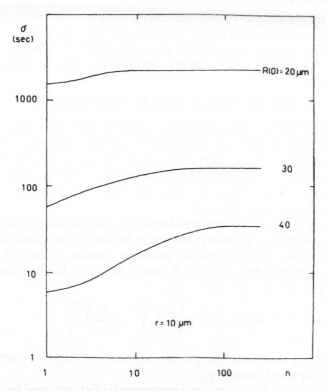

FIG. 8.9. The standard deviation σ of time required to make n captures, for $r = 10\,\mu\text{m}$ and $R(0) - 20$, 30, and 40 μm. Cloud liquid water content $M = 1$ g/m^3. (From Robertson, 1974.)

Statistical growth: the stochastic coalescence equation

In the analyses of Bowen, Telford, and Robertson droplet growth starts with a population consisting of distinct collected droplets and collector drops. What must happen in nature is that a continuous spectrum of droplet sizes, formed by the condensation-diffusion process, evolves by random collisions (at first very rare) and thereby extends itself in the direction of increasing drop size. To be understood, the rain-forming process must therefore be viewed as the evolution of an entire droplet spectrum rather than as the growth of a subset of drops, assumed to be the "collectors" from the start.

To derive the differential equation describing the development in time of the droplet spectrum, we begin by defining the *coagulation coefficient* that describes the likelihood that a drop of radius R will overtake and collide with a droplet of radius r. Suppose these two drops are contained in a unit volume of air. In a unit time, the larger drop sweeps out a volume given by $\pi R^2 u(R)$ and the droplet sweeps out volume $\pi r^2 u(r)$. If the drop is to capture the droplet in unit time, neglecting aerodynamic effects, we see that they cannot be far apart initially and that, in fact, they must be contained in a common volume given by $\pi(R + r)^2[u(R) - u(r)]$. Allowing for aerodynamic effects reduces the size of this common volume to

$$K(R, r) = \pi(R + r)^2|u(R) - u(r)|E(R, r), \tag{8.18}$$

where $E(R, r)$ is the collision efficiency. The quantity $K(R, r)$ is called the coagulation coefficient. The continuous growth equations (8.12) and (8.13) could have been formulated in terms of K. In the stochastic equations this coefficient is interpreted as the *probability* that a drop of radius R collects a droplet of radius r in unit time, given that both are present with unit concentration.

While it is natural to define the coagulation coefficient in terms of drop radii, the stochastic equations turn out to be simpler in appearance if it is transformed to a function of drop volumes. Letting V and v denote the volumes corresponding to radii R and r, we have for the coagulation coefficient

$$H(V, v) = K\left[\left(\frac{3V}{4\pi}\right)^{1/3}, \left(\frac{3v}{4\pi}\right)^{1/3}\right]. \tag{8.19}$$

Thus $H(V, v)$ describes the probability that a drop of volume V will collect a droplet of volume v.

Now we suppose that the drop spectrum is characterized by $n(v)$ such that $n(v)dv$ is the average number of drops per unit volume of space whose volumes are between v and $v + dv$. The total number of coalescences per unit time experienced by drops within this size interval is

$$n(v)dv \int_0^\infty H(V, v)n(V)dV.$$

This integration accounts for all possible captures of the drops in dv by larger drops $(v < V < \infty)$, as well as all captures of smaller droplets by the drops in $dv(0 < V \leqslant v)$. These coalescence events *reduce* the number of drops in dv. But the number of drops in this size interval is *increased* by coalescences between all pairs of smaller drops whose volumes sum to v. This rate of increase is given by

$$\frac{1}{2} dv \int_0^v H(\delta, u)n(\delta)n(u)du,$$

where $\delta = v - u$. The $\frac{1}{2}$ factor is necessary to prevent any particular capture combination from being counted twice.

Taking into account both effects, we have for the rate of change of drop concentration in the size interval dv

$$\frac{\partial}{\partial t} n(v)dv = \frac{1}{2} dv \int_0^v H(\delta, u)n(\delta)n(u)du$$

$$- n(v)dv \int_0^\infty H(V, v)n(V)dV. \qquad (8.20)$$

Variously called the kinetic equation, coagulation equation, or stochastic coalescence equation, this result dates back to Smoluchowski in 1916 according to Drake (1972a) in a comprehensive review. It was first employed in the analysis of rain formation by Melzak and Hitschfeld (1953). Although the formulation of the problem was correct, this early work was impaired for two reasons: the collision efficiencies for small drops, on which the results crucially depend, were not accurately known; moreover, since a high speed computer was not available, the integrations had to be done by hand.

It was a surprisingly long time after computers became generally available that this approach was again undertaken—in this instance by Twomey (1964, 1966). Twomey, motivated by Telford's work described above, apparently rederived the stochastic equation, claiming that this approach embodied, in a more general form, the idea of the statistically "fortunate" drops. It was argued from the first (Warshaw, 1967) that this interpretation of (8.20) is incorrect, partly because the equation is based on the mean drop-size distribution $n(v)$. Controversy continued on whether the equation actually embodies the statistical effects ascribed to it. Theoretical doubts were aggravated by the fact that early results on stochastic coalescence were often in disagreement, owing to numerical errors that arise in the numerical integration of (8.20). By experimenting with different combinations of integration and interpolation schemes, Berry and Reinhardt (1974) established numerical procedures for integrating (8.20) which are very accurate, at least so long as the initial distribution is not too peaked. On the theoretical side there have been contributions by Scott (1968, 1972), Long (1971), Drake (1972b), and Gillespie (1972, 1975). Although there remain some rather subtle unresolved points, it appears from Gillespie's analysis that (8.20) does include that intended statistical effects.

The solution of (8.20) is $n(v, t)$, the droplet spectrum at time t which evolves by coalescence from a given initial distribution $n(v, 0)$. Since in nature the growth is stochastic, it must be recognized that a given distribution $n(v, 0)$ leads to an entire family of solutions $n(v, t)$, one for each "realization" of the coalescence process. The deterministic solution provided by (8.20) corresponds to the average value of $n(v, t)$ over many such realizations. Gillespie showed that, subject to reasonable assumptions, the statistical fluctuations about $n(v, t)$ are Poisson-distributed with parameter n. Consequently the statistical dispersion of $n(v, t)$ diminishes as $1/\sqrt{n}$.

To illustrate the results of integrating (8.20), we give several examples from the study of Berry and Reinhardt (1974a, b). They characterized the droplet spectrum by the density function $f(x)$, where $f(x)dx$ is the number of droplets per unit volume of air in the size interval $(x, x + dx)$, and x stands for droplet mass. Mass and volume are related by $x = \varrho_L v$, and the density functions $f(x)$ and $n(v)$ are related by $f(x)dx = n(v)dv$. Berry and Reinhardt also considered the mass density function $g(x)$, defined by $g(x) = xf(x)$. The droplet concentration N and liquid water content L are expressed as integrals over the density functions:

$$N = \int f(x)dx$$

$$L = \int g(x)dx.$$

The mean mass of a droplet in this system of equations is given by

$$x_f = \int xf(x)dx \bigg/ \int f(x)dx = L/N.$$

A droplet whose size equals the mean droplet mass has a radius given by

$$r_f = (3/4\pi\varrho_L)^{1/3}x_f^{1/3}.$$

The mean mass of the mass density function is defined by

$$x_g = \int xg(x)dx \bigg/ \int g(x)dx$$

$$= \int x^2 f(x)dx \bigg/ \int xf(x)dx,$$

and its equivalent radius is

$$r_g = (3/4\pi\varrho_L)^{1/3}x_g^{1/3}.$$

Figure 8.10 shows the evolution of a droplet spectrum by coalescence, starting with $f(x)$ in the form of a gamma distribution with $r_f = 12\,\mu$m, a liquid water content of 1 g/m^3, and a concentration of 166 droplets per cm^3. The spread of the distribution is characterized by the relative variance, defined as var $x = (x_g/x_f) - 1$. In this example, the initial value of the relative variance is equal to unity. The function plotted is the log-increment mass density function $g_l(r)$, defined such that $g_l(r)d(\ln r)$ equals the mass of cloud droplets per unit volume of air with radii in $d(\ln r)$. This distribution emphasizes the large droplets, and is related to the fundamental density function by $g_l(r) = 3x^2 f(x)$. The initial unimodal distribution is seen to evolve systematically by coalescence into a bimodal distribution. The dashed lines indicate the development in time of r_f and

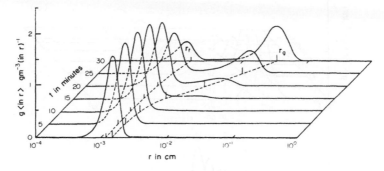

FIG. 8.10. Example of the development of a droplet spectrum by stochastic coalescence. (From Berry and Reinhardt, 1974b.)

r_g. It is evident that the first mode follows r_f closely, and the second mode is approximated by r_g.

To clarify the significance of different physical processes in broadening the distribution, Fig. 8.11 illustrates the growth of an initial distribution consisting of two modes, a spectrum S_1 centered at 10 μm with a water content of 0.8 g/m^3 and a spectrum S_2 centered at 20 μm with a water content of 0.2 g/m^3. Results of calculations are shown for four different assumptions: (a) collision between all droplet pairs is allowed; (b) collision is permitted only for droplets in S_1; (c) collision can only take place between a drop in S_2 and a droplet in S_1; (d) collision is restricted to the drops in S_2. Berry and Reinhardt refer to processes (b), (c), and (d) as autoconversion, accretion, and large hydrometeor self-interaction.

The results for case (a) are similar to those in Fig. 8.10. The spectrum S_1 is depleted and its mode does not increase much from the initial value. The amount of water in S_2 increases and the mode of this distribution is closely approximated by r_g. In case (b), S_2 gains only by the interaction of the droplets in S_1, and the effect is small. The rate of transfer of water to S_2 is more rapid in (c), showing that accretion is more efficient than autoconversion in transferring water from small to larger drops. Case (d) shows that it is the interactions between large drops that account for a flattening of the tail of the distribution and its extension to larger sizes.

These results demonstrate three basic modes of collection that operate to produce large drops. Autoconversion serves to add water to S_2 initially so that other modes can operate. Its rate, however, is generally slower than the other processes. Accretion is the main mechanism for transferring water from S_1 to S_2. S_1 is depleted uniformly, maintaining its general shape and position, eventually losing almost all of its water to S_2. The third mode, large hydrometeor self-collection, produces large drops quickly. It is responsible for the rapid increase of the predominant radius

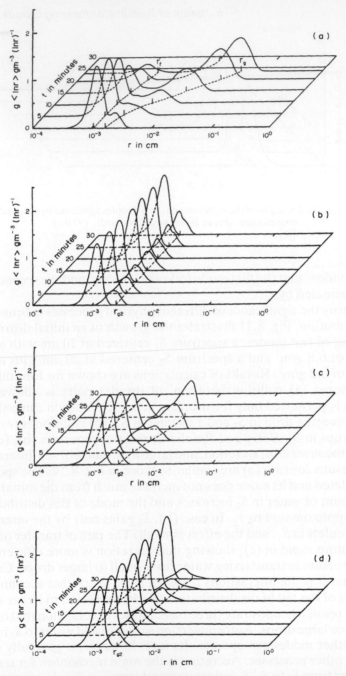

FIG. 8.11. Comparison of the evolution by coalescence of a drop spectrum, initially bimodal, by different collection processes. In each case the initial distribution consists of a spectrum of droplets S_1 centered at $10\,\mu m$ radius and a spectrum S_2 of drops centered at $20\,\mu m$. (a) All collisions accounted for. (b) Only collisions between droplets in S_1 allowed. (c) Only collisions between a drop in S_2 and a droplet in S_1 allowed. (d) Only collisions between drops in S_2 considered. (Adapted from Berry and Reinhardt, 1974a.)

r_g and the emerging shape of S_2. The rate of this process increases as water mass is added to S_2 by accretion.

Condensation plus stochastic coalescence

East (1957) made calculations of droplet development by condensation and coalescence in moderate updrafts of several meters per second, finding that precipitation-size drops could be produced in realistically short periods of time. The stochastic coalescence equation was not actually used, the droplet population from the start being separated into collector drops and collected droplets. Also the collision efficiencies were inaccurately known. East's main contribution was to explain how condensation, which causes the droplet spectrum to narrow, actually speeds up the coalescence process. The spectrum narrows by condensation because small droplets grow (in radius) faster than large droplets. As the small droplets grow, their collision efficiency relative to the large drops increases at a rapid rate (see Fig. 8.3), and coalescence is thus accelerated. The relative velocity between large and small droplets, on which coalescence also depends, remains essentially unchanged in this narrowing process. Considering any two droplets of different initial size, we see from the approximate equation (7.19) that the difference between the squares of the radii remains constant. For droplet sizes less than about 40 μm the fall speed is given by Stokes' Law and the relative velocity between the droplets is constant. The net effect of condensation-narrowing is therefore to accelerate coalescence.

More exact calculations of droplet evolution by condensation and coalescence were carried out by Ryan (1974). Condensation was modeled using (7.40) and assuming an updraft speed of 1 m/s. The initial drop spectrum was a gamma distribution with a concentration of 200 cm^{-3}, a liquid water content of 1 g/m^3, and a dispersion (ratio of standard deviation of radius to mean radius) of 0.2. The collision efficiencies of Klett and Davis (1973) were used, and the temperature lapse rate was assumed to be moist adiabatic.

Figure 8.12 compares the cases with and without condensation. Coalescence by itself is only able to produce drops smaller than 60 μm after an elapsed time of 1000 s. With condensation there is a tendency for the spectrum to become narrower over the size range below 25 μm. However, the general shift of these droplets to larger sizes by condensation increases their collision efficiency and promotes more rapid growth by collection. The creation of a significant number of drizzle-size drops by 1000 s clearly indicates the importance of condensation in accelerating coalescence.

Calculations similar to those of Ryan, but with fewer approximations, were reported by Ochs (1978). Starting with an activity spectrum of condensation nuclei, he calculated the development of the droplet

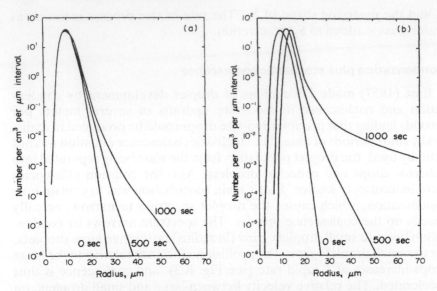

FIG. 8.12. The effect of condensation on growth by coalescence. The same droplet spectrum evolves by coalescence, (a) without, and (b) with, an allowance for condensation. (Adapted from Ryan, 1974.)

spectrum in an ascending air parcel by condensation and coalescence, specifically accounting for the supersaturation. Figure 8.13 is an example of the results, for conditions appropriate for a maritime cloud that forms on nuclei with an activity spectrum given by N_c (in cm^{-3}) $= 105\ s^{0.63}$.

The supersaturation rises rapidly and nuclei are activated. After the droplet concentration reaches a maximum the supersaturation falls to a

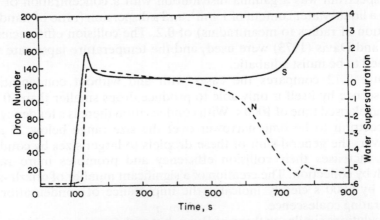

FIG. 8.13. Time histories of supersaturation (%) and droplet concentration (cm^{-3}) in a cloud formed on a maritime distribution of condensation nuclei. (Adapted from Ochs, 1978.)

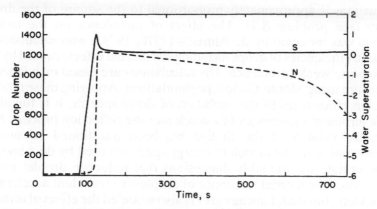

FIG. 8.14. As Fig. 8.13, except for a continental distribution of condensation nuclei. (Adapted from Ochs, 1978.)

steady value. The droplets are then depleting the water vapor at the same
rate it is produced by the updraft. As the droplets grow, their collection
efficiencies increase, setting the stage for coalescence. Once it begins,
coalescence proceeds rapidly, as indicated by the fast decline in the
number of drops. At the same time, the supersaturation increases
sharply because the drops, now fewer in number, are no longer able to
consume the excess vapor at the rate it is created. The increasing
supersaturation activates new condensation nuclei, at first causing an
increase in drop number. This is only a transient effect, however,
because the newly-formed droplets are quickly consumed by coalescence. Figure 8.14 is for a similar calculation for a continental cloud, with
activity spectrum $N_c = 1450\ s^{0.84}$. The onset of coalescence is signalled
once again by the reduction in drop concentration, but in this cloud the
numbers remain great enough to prevent a sudden rise in supersaturation
and the activation of new nuclei. In his calculations, Ochs allowed for the
collision-induced breakup of large drops. This effect is important in
establishing the distribution of raindrops with size, and is described in
Chapter 10.

The effects of turbulence on collisions and coalescence

Some degree of turbulence is always present in clouds. Intuitively, it
might be expected that turbulent fluctuations would promote coalescence by increasing the chances of droplet collisions.

Although it has proven to be very difficult to predict the effects of
turbulence with certainty, two mechanisms are recognized by which
turbulence may affect the collection process. The first is based on the fact
that drops of different sizes respond differently to a fluctuating velocity
field. For small droplets, the relaxation time in response to velocity

fluctuations is approximately proportional to the square of the droplet radius (see problem 8.7). The effect of turbulence on the collision process was modeled by de Almeida (1976, 1979), who calculated the collision efficiencies of drops including the inertial effects caused by their response to weak turbulence. His calculations are based on a laborious procedure using Monte-Carlo type simulations. Applying these modified collision efficiencies to the evolution of cloud spectra, it is found that weak turbulent motions greatly accelerate the collection process. However, the validity of this finding has been questioned because the representation of the turbulent energy spectrum used by de Almeida is inappropriate for the eddy dimensions that influence droplet interactions. Also, the inertial resistance of droplets to turbulent accelerations was underestimated. Panchev (1971) also modeled the effect of turbulent inertia. He found that the turbulence-induced relative velocities between drops of different size were small compared to the difference in their terminal fall speeds. Gravitational coalescence seems thus to be more important than any coalescence induced by differential response to velocity fluctuations.

The second mechanism that can influence coalescence in a turbulent medium may be important for drops having the same size or different sizes. This mechanism is related to the eddy structure of the turbulent velocity field. At a given time, cloud drops can be contained in two spatially separated turbulent eddies. Later, it might happen that these eddies overlap, causing collisions and coalescence of some of the drops. Reuter *et al.* (1988) formulated a model that attempts to estimate the probability of drop collisions caused by overlapping eddies. For given values of drop radii and turbulence intensity, the model estimates the probabilistic collection kernel. The model is based on a set of stochastic differential equations that govern the relative motion of the drops. The magnitude of the random perturbations is related to the coefficient of turbulent diffusion. Results indicate that as the rate of diffusion increases, so does the magnitude of the collection kernel. To determine the overall effect of turbulence on the evolution of a droplet spectrum, the collection equation (8.20) must be solved including the new probabilistic kernel. Results of this approach suggest that the overlapping of turbulent eddies contributes to the collection process and that the importance of this effect increases with increasing intensity of turbulence, but that the influence on droplet growth rate is nevertheless small.

An important property of turbulence that has not yet been considered in calculations of droplet collisions is its intermittent character. This intermittence manifests itself by abrupt inhomogenieties in the velocity field, with strong energy dissipation in some regions and very little in others. Because the effects of turbulence on droplet collisions are likely to be highly nonlinear, a turbulence field consisting of a few patches of intense turbulence surrounded by regions of weak turbulence may have a more significant influence on droplet growth than a field of homogeneous turbulence having the same average intensity.

In summary, it must be stated that the effect of turbulence on coalescence is not well understood. Preliminary results for a homogeneous turbulence field seem to indicate that the collection of droplets is promoted by the differential response to turbulence and by the overlapping of turbulent eddies. Further work is needed to substantiate these findings and to quantify the relationships between the collection kernel and the turbulence intensity. Also, the effects of intermittent turbulence must be investigated before final conclusions can be drawn.

Concluding remarks

Observations show that rain can develop in warm clouds of the cumulus type in times as short as about 15 min after the cloud begins to form. It is generally agreed that the process responsible for this development must be gravitational coalescence among the droplets, so that the droplet populations most likely to produce rain in a short time are those with broad enough spectra to have a high rate of collisions. A serious impediment to coalescence growth is the fact that collection efficiencies between small droplets are extremely small. Moreover, the condensation-diffusion process, which dominates droplet growth initially, leads to a narrowing of the spectrum, seemingly complicating the coalescence problem. The theoretical task is therefore to explain raindrop development in reasonable times in spite of the small collection efficiencies and the approximate parabolic form of the diffusional growth law.

It is now recognized that statistical effects are crucial in the early stages of coalescence, so that the stochastic equation should be used to describe this process. Even so, coalescence alone is not sufficient to account for rain development in reasonably short times, starting with realistic droplet spectra. Several mechanisms have been postulated for broadening the spectrum, thus paving the way for rapid coalescence. There is no agreement on which of these is the most important, although each may be effective under certain conditions. Rather than spectral broadening, it has been shown that the additional effect which can account for rain development, at least in some conditions, is continued condensation growth. This seems strange at first since condensation-diffusion, even with the kinetic corrections, leads to a narrowing of the droplet spectrum. Though the spectrum narrows the rate of coalescences increases because the collision efficiency increases rapidly as the droplets in the spectrum grow. Definitive calculations have not yet been made which account for rain development under various conditions of updraft speed, temperature, initial droplet spectrum, and entrainment mixing. Results from simple models are encouraging, however, indicating that theory may be on the verge of explaining the observations.

Problems

8.1. On a particular day the orographic cloud on the island of Hawaii is 2 km thick with a uniform liquid water content of 0.5 g/m^3. A drop of 0.1 mm radius at cloud top begins to fall through the cloud.

   (a) Find the size of the drop as it emerges from cloud base, neglecting vertical air motions in the cloud. In this and subsequent parts of the problem, neglect growth by condensation, use the elementary form of the continuous-growth equation, and assume a collection efficiency of unity.

   (b) Assuming that the terminal velocity of the drop is equal to $k_3 r$, where $k_3 = 8 \times 10^3$ s^{-1}, find the time taken by the drop to fall through the cloud.

   (c) Hawaiian orographic clouds are actually maintained by gentle upslope motions, which cause a steady, weak updraft. Solve for the size of the drop in part (a) as it emerges from the cloud if there is a uniform updraft of 20 cm/s.

8.2. To allow for ventilation effects, the diffusional growth equation is sometimes approximated by

$$r \frac{dr}{dt} = (S - 1)\xi_1 (1 + 0.3Re^{1/2}),$$

where $\xi_1 = [F_k + F_d]^{-1}$ and Re is the Reynolds number characterizing the flow around the drop of radius r. Assess the importance of ventilation by comparing the time required for a drop to evaporate completely using (1) the approximation including the ventilation factor and (2) the same equation neglecting this factor. Make the comparison for drops with initial radii of 1 mm, 0.5 mm, and 0.1 mm. For environmental conditions assume $T = 10°$C, $p = 80$ kPa, and a relative humidity of 50%. Approximate the drop fall velocity by the linear law of problem 8.1.

8.3. Neglecting ventilation effects, calculate the distance a drop of 0.1 mm radius falls while evaporating completely under the conditions of problem 8.2. Compare the solution based on the linear fall speed approximation with a solution using the data of Gunn and Kinzer (Table 8.1).

8.4. A drop of 0.2 mm diameter is inserted in the base of a cumulus cloud that has a uniform liquid water content of 1.5 g/m^3 and a constant updraft of 4 m/s. Using the elementary form of the continuous-growth equation and neglecting growth by condensation, determine the following:

   (a) the size of the drop at the top of its trajectory;
   (b) the size of the drop as it leaves the cloud;
   (c) the time the drop resides in the cloud.

   Assume a collection efficiency of unity, and for the dependence of fall velocity on size use the data in Table 8.1. (Note: Parts (a) and (b) of this problem are well suited for graphical solution.)

8.5. One member of a population of cloud droplets, all 10 μm in radius, grows by condensation and coalescence. Assume a cloud water content of 1 g/m^3 and a supersaturation of 0.2%, and solve for the time it takes for the drop to reach 20, 30, and 40 μm radius. For simplicity, assume that the droplets remain 10 μm in size. Use the continuous-growth equation for coalescence, allowing for the size and fall speed of the droplets. For the fall speed use Stokes' Law. Allow for the collection efficiency of the drop relative to the droplets using the data in Table 8.2, and assuming a coalescence efficiency of unity. This problem should be solved graphically.

8.6. A small drizzle drop is swept upwards in a cumulus congestus cloud and grows by accretion and condensation in the supersaturated environment. The condensation parameter ξ may be regarded as constant and the linear fall speed law of problem 8.1 approximates the relative velocity between the growing drop and the cloud droplets. Develop and solve the differential equation that describes the growth of the drop by accretion and condensation acting simultaneously. Compare the result with the approximation obtained by adding the solutions for growth by accretion and by condensation acting separately.

8.7. Prove that cloud droplets move very quickly in response to changes in the ambient air velocity, by solving for the effective time constant characterizing the response. Determine the relationship between this time constant and the droplet terminal fall velocity.

9

Formation and Growth of Ice Crystals

ONCE a cloud extends to altitudes where the temperature is colder than 0°C ice crystals may form. Two phase transitions can lead to ice formation: the freezing of a liquid droplet or the direct deposition (sublimation) of vapor to the solid phase. Both are nucleation processes, and in principle homogeneous and heterogeneous nucleation are possible.

A newly formed ice crystal in a cloud of liquid droplets is in a favorable environment to grow rapidly by diffusion. The vapor in the cloud is essentially saturated relative to liquid water and hence supersaturated relative to ice. In a time of only a few minutes, such a crystal can grow to a size of many tens of micrometers in linear dimension. A crystal this size falls with a velocity of a few tens of centimeters per second. It may eventually reach the ground as an individual crystal, or it may collide with supercooled droplets to form a rimed crystal, or with other crystals to form a crystal aggregate or snowflake.

The growth processes are thus the same as for a droplet, namely diffusion followed by coagulation. For the crystal, however, diffusional growth is more significant than for the droplet because of the difference in saturation vapor pressure over water and ice.

Nucleation of the ice phase

It is convenient to consider first the homogeneous phase transitions that can lead to the formation of ice. Homogeneous freezing of a pure liquid drop occurs when statistical fluctuations of the molecular arrangement of the water produce a stable, icelike structure that can serve as an ice nucleus. Just as for the homogeneous nucleation of liquid in the vapor, discussed in Chapter 6, two considerations determine the conditions for homogeneous nucleation of freezing: the size of the stable nucleus and the probability of occurrence of embryonic ice nuclei by random rearrangement of water molecules. These quantities depend on the surface free energy of a crystal/liquid interface, which is analogous to the surface tension at a liquid/vapor interface. The numerical value of the surface free energy is not known accurately, but is accepted to be close to 2.0×10^{-2} N/m (20 erg/cm^2). Consistent with experimental data on the freezing of pure water, this value inserted in the equations predicts that droplets smaller than 5 μm will freeze spontaneously at a temperature of about $-40°$C. Larger droplets are predicted to freeze at slightly warmer temperatures, also in agreement with most observations. In natural clouds, few liquid drops are thought ordinarily to exist at temperatures as cold as $-40°$C, implying that heterogeneous freezing has occurred at temperatures warmer than $-40°$C. However, occasional reports of aircraft icing in clouds at temperatures near $-40°$C suggest that droplets sometimes may not freeze until the homogeneous threshold is reached.

Some liquid in clouds as cold as $-20°$C is not at all uncommon. This resistance to freezing is contrary to common experience, which indicates that water freezes when the temperature falls below 0°C. Our experience is based on observations of bulk water, in which a single nucleation event anywhere suffices to cause the entire mass to freeze. A cloud is an unusual system in which the water mass is distributed over a large number of very small drops, each one of which must experience a nucleation event before the cloud can be entirely frozen.

Homogeneous deposition occurs when vapor molecules form a stable ice embryo by chance collisions. Although the surface free energy of a crystal/vapor interface is poorly known, theory predicts that homogeneous nucleation of deposition should only occur for extreme conditions of supersaturation. More than twenty-fold supersaturation with respect to ice is required at a temperature a few degrees below 0°C, and still higher supersaturations at colder temperatures. Experimental confirmation of the theory of homogeneous deposition seems out of the question because liquid water droplets should always nucleate homogeneously before the supersaturation reaches the high values required for ice. These droplets will freeze spontaneously at temperatures colder than $-40°$C, making it impossible to recognize ice crystals that might be formed by deposition. In spite of the experimental uncertainties, it is clear that homogeneous deposition cannot occur in the atmosphere, where the necessary extreme supersaturations never exist.

Ice crystals usually appear in a cloud in appreciable numbers when the temperature drops below about $-15°$C, signifying heterogeneous nucleation. Water in contact with most materials freezes at temperatures warmer than $-40°$C, and deposition occurs on most surfaces at supersaturations and supercoolings less than the homogeneous nucleation values. Thus the nucleation of ice in supercooled water or a supersaturated environment is aided by the presence of foreign surfaces or suspended particles.

The foreign material provides a surface or *substrate* on which water molecules can impinge, stick, bond together, and form aggregates with an icelike structure. The larger the aggregate, the more likely it is to be stable and continue to exist. The probability of heterogeneous nucleation of freezing or deposition depends strongly on the properties of the substrate material as well as on the supercooling and supersaturation. The more tightly-bound the water molecules are to the substrate, the greater will be the probability of ice nucleation. In addition, if the crystal structure of the substrate closely resembles that of an ice crystal plane, it will increase the chances of ice nucleation. When the binding and the matching of the crystal lattice are good, the supersaturation or supercooling required to nucleate ice on a substrate may be much lower than that for homogeneous ice nucleation.

Supercooled clouds in the atmosphere develop and exist in the presence of vast numbers of aerosol particles, a small fraction of which serve as ice nuclei at temperatures considerably warmer than the $-40°C$ threshold for homogeneous freezing. Several nucleation mechanisms are possible, and are shown schematically in Fig. 9.1. Ice may form directly from the vapor phase on suitable deposition nuclei. Three modes of activation are recognized for freezing nuclei. Some serve first as centers for condensation, then as freezing nuclei. Some promote freezing at the instant they come into contact with a supercooled droplet. Others cause freezing after becoming embedded in a droplet. A given particle might nucleate ice in different ways, depending on the ambient conditions and its history in the cloud. The relative importance of the different freezing

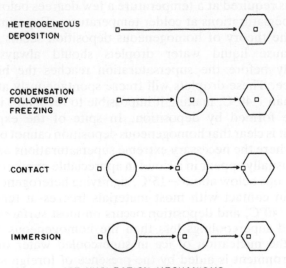

ICE NUCLEATION MECHANISMS

FIG. 9.1. Schematic picture of the different ways atmospheric ice nuclei can account for ice formation.

modes in the atmosphere has not been established. It is also difficult to distinguish between deposition and freezing nucleation when ice nucleates on an insoluble surface in an environment supersaturated relative to liquid water. Even in conditions below water saturation nucleation need not imply deposition, because the nuclei may contain soluble components. The soluble material may nucleate the liquid phase below water saturation and enable the insoluble material to nucleate ice by freezing. Because of the confusion between mechanisms, one often speaks of "ice nucleation" as the phenomenon instead of the more specific "freezing nucleation" or "deposition nucleation". Likewise, the atmospheric particles serving as nucleation centers can most safely be referred to as "ice nuclei".

It is clear that the subject of ice nucleation is beset with theoretical and experimental uncertainties. Knight (1979) may be consulted for a review of some of the fundamental questions.

Experiments on heterogeneous ice nucleation

The nucleating properties of small particles are studied by introducing the particles into cloud chambers with controlled supercooling and supersaturation. The conditions are noted where the onset of nucleation occurs. (Ice crystals are usually discernable even in the presence of water cloud by the scintillation of light scattered from a strong beam.) In these experiments it is usually not possible to distinguish between a deposition event and a condensation event followed by freezing. Other experiments consist of adding finely divided material to supercooled, purified water and noting the threshold temperature for freezing. Table 9.1 summarizes the threshold ice-forming temperatures of certain pure and natural substances. Also listed are the dimensions of the two axes that characterize the lattice structure of the crystalline substances. The material matching ice the closest in structure is silver iodide, which also has a relatively warm nucleating temperature. It is this property of silver iodide, and the finding that the material can be efficiently generated in the form of extremely fine particles by burning special compounds of silver, that has made AgI the most widely used substance for artificial cloud seeding. The table shows that lattice dimensions are not the only factor determining nucleating ability. Especially noteworthy are the warm nucleation thresholds of several organic materials that do not have a well-defined crystal structure. Some combination of lattice matching, molecular binding, and low interfacial energy with ice accounts for the nucleating ability of a substance. Theory and experimental methods are not yet able to explain the relative importance of these factors and to enable prediction of the nucleating properties of a substance from fundamental principles.

TABLE 9.1. *Temperatures at which different substances nucleate ice. (From Houghton, 1985)*

Substance	Crystal lattice dimension		Temperature to nucleate ice (°C)	Comments
	a axis (Å)	*c* axis (Å)		
Pure substances				
Ice	4.52	7.36	0	—
AgI	4.58	7.49	−4	Insoluble
PbI$_2$	4.54	6.86	−6	Slightly soluble
CuS	3.80	16.43	−7	Insoluble
CuO	4.65	5.11	−7	Insoluble
HgI$_2$	4.36	12.34	−8	Insoluble
Ag$_2$S	4.20	9.50	−8	Insoluble
CdI$_2$	4.24	6.84	−12	Soluble
I$_2$	4.78	9.77	−12	Soluble
Minerals				
Vaterite	4.12	8.56	−7	
Kaolinite	5.16	7.38	−9	(Silicate)
Volcanic ash	—	—	−13	
Halloysite	5.16	10.1	−13	
Vermiculite	5.34	28.9	−15	
Cinnabar	4.14	9.49	−16	
Organic materials				
Testosterone	14.73	11.01	−2	
Chloresterol	14.0	37.8	−2	
Metaldehyde	—	—	−5	
β-Naphthol	8.09	17.8	−8.5	
Phloroglucinol	—	—	−9.4	
Bacterium	—	—	−2.6	(Bacteria in leaf mold)
Pseudomonas Syringae				

Atmospheric ice nuclei

Several methods have been used to study atmospheric ice nuclei, the most common of which are cloud chambers, and filter systems, into which samples of air are drawn. In the cloud chambers the sample is cooled down to a controlled temperature and a cloud is formed by adding sufficient water vapor. An optical system or some other means of ice crystal detection is used to count the number of crystals that form as a function of the degree of supercooling. This kind of experiment is not specific as to the mode of nucleation and gives no information about nucleus size unless the incoming air sample is filtered to remove particles larger than a preselected size.

The second method consists of collecting aerosols by drawing the air sample through filter paper with known pore sizes. The particulates trapped on the filters are then introduced to an environment suitable for ice crystal development and observations are made of the number of crystals that form on a substrate. This technique does give information about the size of the nuclei but not about their mode of activation.

The concentrations of atmospheric ice nuclei inferred from these methods of detection are subject to uncertainty because of the major influence of the history of the aerosol, the humidity of the cloud chamber, and other experimental variables. Reported concentrations at temperatures ranging from $-15°C$ to $-20°C$ extend over many orders of magnitude. As the supercooling increases, so does the nucleus count at usually a rapid rate. Ice nucleus concentrations also increase with increasing supersaturation, which quantity is not always controlled or measured in experiments. Furthermore, there is evidence that some nucleation events do not occur immediately, but require a long exposure of the nucleus to supercooled conditions. Accordingly, only a fraction of the nuclei actually present in an air sample may be activated during the time of an experiment. Although much of the apparent variability in ice nucleus counts may be explained by experimental limitations, it is clear nevertheless that the ice nucleus content of the air is a highly variable quantity. Fletcher (1962), reviewing the data to that time, gave as a typical concentration one nucleus per liter of air at a temperature of $-20°C$, increasing by a factor of ten for each 4°C of additional cooling. This exponential dependence of concentration on supercooling is still accepted as typical, but it is recognized that the count at any given location and time can be at least an order of magnitude above or below the typical relationship.

Taking 10^4 cm^{-3} as a typical concentration of aerosol particles, we see that the one nucleus per liter active at $-20°C$ is only one particle in 10^7. Separating and identifying such rare particles is not an easy task. Although uncertainties remain, much evidence points to clay minerals, especially kaolinite (see Table 9.1), as a major component of atmospheric ice nuclei. This is a common material found in many soil types. The nucleation threshold is a fairly warm $-9°C$. Snowflakes collected as they fall to the ground are usually found to contain particles that appear to be centers for crystal growth. These particles have been identified by electron microscopy to be kaolinite with sizes ranging from 0.1 μm to 4 μm. What is not clear is how kaolinite could explain the occurrence of ice in clouds warmer than $-9°C$, as sometimes observed.

Another source of ice nuclei has been revealed by the discovery that the bacteria in decaying plant leaf material can be effective nucleants at warm temperatures (Schnell and Vali, 1976; Vali *et al.*, 1976). It had already been known that common soil particles could be active at temperatures warmer than the threshold for kaolinite, possibly explained by submicron-size nuclei of some minor organic substance. The work by Vali and his colleagues showed that the bacterium *Pseudomonas syringae* itself serves as an ice nucleus at a temperature as warm as −1.3°C, although its nucleating ability is a rare and changeable property. The overall significance of biogenic nuclei in the atmosphere has not been established, and remains a subject in need of more research.

Meteoric material has been thought to be a possible source of atmospheric ice nuclei, initially because of an apparent correlation found by E. G. Bowen between extreme rainfall occurrences and meteor showers, and more recently because of the demonstration that submicron meteorites, produced by vaporizing and recondensing meteoritic material, are rather effective ice nucleants. From growing observational evidence, however, it appears that terrestrial sources account for most atmospheric ice nuclei. Measurements at coastal sites indicate more nuclei in air from trajectories over land than from over the ocean. The concentrations of nuclei also tend to decrease with altitude over land, consistent with a source at the surface. Even at the South Pole, the particulates in snowflakes are found to be clay minerals.

It seems likely that more than one kind of material may serve as atmospheric ice nuclei. Depending on temperature, humidity, and proximity to sources, one may be more important than the others in a given cloud. Although the atmosphere has an abundance of condensation nuclei, ice nuclei are scarce, regardless of their origin. Accordingly, the supercooling of cloud water to −15°C or colder is not uncommon, although supersaturations exceeding 1% in the atmosphere are extremely rare. Some of the outstanding questions about atmospheric ice nuclei were reviewed by Vali (1985).

The ice phase in clouds

The occurrence of ice crystals in clouds is related to cloud type, temperature, and cloud age. Overall, observations confirm that the colder the cloud temperature, the greater is the likelihood that some ice crystals are present along with the supercooled water droplets. Although essentially no clouds with tops as cold as −20°C are ice-free, ice anywhere in a cloud is unlikely unless the top extends to −5°C or colder. Ice is more common in decaying cumulus clouds than in newly developing clouds, and is probably more common in stratiform clouds than in cumulus clouds with the same cloud-top temperature.

Measured concentrations of ice crystals in clouds range from the lower limit of detectability (often about 0.01 per liter) to about 100 L^{-1}. Concentrations are high in cirrus clouds, and still higher in ice fogs, which develop under extremely cold arctic conditions. The most perplexing question in ice microphysics is to explain the vast discrepancies, some-

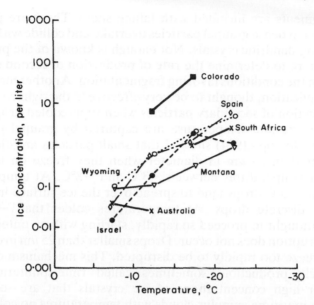

FIG. 9.2. Summary of observed ice particle concentrations for clouds in which secondary processes of ice formation are believed to be unimportant. Temperatures refer to the origins of the ice particles; the data have been averaged over intervals of 5°C. (Data compiled by W. A. Cooper.) (Adapted from Vali, 1985.)

times amounting to a factor of 10^4, between observed crystal concentrations and the measured concentrations of ice nuclei in the ambient air.

The first crystals to appear in a cloud must form on ice nuclei. (An exception to this generalization may be cirrus clouds, which form at such cold temperatures that homogeneous freezing may occur as soon as the liquid phase appears.) Additional crystals may then be produced by secondary processes in which the primary crystals are "multiplied". Figure 9.2 shows the observed ice particle concentrations in clouds in which secondary processes of ice formation are thought to be unimportant. For such clouds there is approximate agreement between ice crystal concentrations and ice nucleus counts in the air below cloud base. Mossop (1985) explained that ice concentrations usually agree with the nucleus concentrations to within a factor of ten in clouds that contain no drops larger than 25 μm diameter at temperatures warmer than −10°C. Other cloud types have been observed to contain crystals far in excess of the ice nucleus counts, with greater discrepancies than can be explained by uncertainties in the measurements.

Two mechanisms are recognized as contributing to secondary ice particle production: the fracture of ice crystals and the shattering or splintering of freezing drops. It is a common observation that many crystal fragments are included with falling snow. These are probably produced when dense graupel particles overtake and collide with fragile, slower-falling dendritic crystals. Not enough is known of the process of crystal fracture to determine the rate of production of secondary particles, or even the conditions favoring fragmentation. Another mechanism of ice multiplication, thought to be very effective in the right conditions, is the production of secondary particles when supercooled drops of the appropriate size and temperature are captured by graupel particles. Hallett and Mossop (1974) showed that small particles are ejected by drops larger than 25 μm in diameter when they freeze on to an ice substrate at temperatures between $-3°C$ and $-8°C$. At temperatures warmer than $-3°C$ drops tend to spread over the ice surface instead of freezing as discrete drops. At temperatures colder than $-8°C$ the freezing is thought to proceed so rapidly, starting with an outer shell of ice, that disruption does not occur. Drops smaller than 25 μm in diameter probably freeze too rapidly to be disrupted. This mechanism of secondary particle production, sometimes called rime-splintering, may account for high concentrations of ice crystals that are sometimes observed in maritime cumulus clouds with temperatures no colder than $-10°C$.

There is no doubt that ice crystals are produced by secondary processes, but because our understanding of these processes is very limited, many observations of crystal concentrations cannot be explained quantitatively (Mossop, 1985).

Diffusional growth of ice crystals

When the first ice crystals nucleate in a cloud, they find themselves in an environment in which the vapor pressure is equal to or slightly greater than the equilibrium vapor pressure e_s over liquid water. The saturation ratio relative to ice may be written

$$S_i = \frac{e}{e_i} = \frac{e}{e_s}\frac{e_s}{e_i} = S(e_s/e_i) \qquad (9.1)$$

where S denotes the saturation ratio with respect to water. The supersaturation ratio, $(e_s/e_i) - 1$, is plotted in Fig. 9.3 from the data in Table 2.1. This shows that a water-saturated cloud has high supersaturation relative to ice and is a favorable environment for rapid growth by diffusion and deposition. The environment will remain favorable for crystal growth as long as liquid droplets are available to evaporate and maintain the vapor pressure at equilibrium relative to water. If the droplets eventually disappear, by evaporation or freezing, the saturation ratio will decrease to equilibrium relative to ice.

The problem of determining the rate of growth of a crystal by diffusion is analogous to that of the growth of a water droplet by condensation but with a complication because of the nonspherical shape of the crystal. Maxwell in his theory of the wet-bulb thermometer solved the equations of heat and mass transfer by drawing on the analogy between the diffusion equation and equations in electrostatics that describe the distribution of potential around a charged conductor. The electrostatic analogy is the starting point of the theory of ice crystal growth by diffusion. By using Poisson's equation in electrostatics and Green's theorem, it can be shown that the integral over the surface of a conductor of the normal component of $-\nabla\phi$, where ϕ is the electrostatic potential, is equal to $4\pi C\phi_s$ where C is the capacitance of the conductor and ϕ_s its potential. If we identify $-D\nabla n$ which is the flux of water molecules with $-\nabla\phi$, then it follows that the total current of water out of the ice crystal is by analogy $4\pi CD(n_s - n_\infty)$ where n_s is the vapor number density at the crystal's surface and n_∞ is its value far from the surface. The assumptions that $\nabla^2 n = 0$ and that n_s is the same over all points on the surface are necessary to complete the analogy. The generalized growth equation therefore becomes

$$\frac{dm}{dt} = 4\pi CD(\varrho_v - \varrho_{vr}). \tag{9.2}$$

C denotes the electrical capacitance, with length units, a function of the size and shape of the particle. For a sphere, $C = r$ and (9.2) reduces to the growth equation for a water droplet. For a circular disk of radius r, which can be used as an approximation for plate-type ice crystals, $C = 2r/\pi$. Ice needles may be approximated by the formula for a prolate spheroid of major and minor semi-axes a and b, for which

$$C = \frac{A}{\ln\left[(a + A)/b\right]},$$

where $A = \sqrt{a^2 - b^2}$. For an oblate spheroid,

$$C = a\varepsilon/\arcsin \varepsilon,$$

where the ellipticity $\varepsilon = \sqrt{1 - b^2/a^2}$.

Actual ice crystals have more complex shapes than the sphere, disk, and ellipsoids for which these theoretical formulas apply. Plane dendrites and plates, however, which are common crystal types, can be reasonably approximated by a circular disk of equal area. Likewise needles can be approximated by long prolates. According to Houghton (1950), who argued from the electrostatic analogy, the field about a plate covered with points is the same as that produced by a smooth plate a suitable distance away. The field is distorted by the irregularities only at distances from the plate which are the same order as the length and spacing of the irregularities. As the water vapor approaches the crystal, the fine structure determines how and where it will be deposited; but this does not affect the total flux of water vapor. Houghton's ideas were confirmed by laboratory measurements (McDonald, 1963b) of the capacitance of brass models of snowflakes.

As the ice crystal grows its surface is heated by the latent heat of sublimation and the value of ϱ_{vr} is raised above the value that would apply without heating. Under stationary growth conditions the value of ϱ_{vr} is determined by the balance between the rates of latent heating and heat transfer away from the surface, which balance is expressed

$$\frac{\varrho_v - \varrho_{vr}}{T_r - T} = \frac{K}{L_s D}. \tag{9.3}$$

Following the thermodynamic arguments in Chapter 7, (9.2) and (9.3) may be combined to give an analytical expression for crystal growth rate. This formula is precisely the same as was obtained for water drops if we replace r by C, $e_s(T)$ by $e_i(T)$ and L by L_s, the latent heat of sublimation:

$$\frac{dm}{dt} = \frac{4\pi C(S_i - 1)}{\left[\left(\frac{L_s}{R_v T} - 1\right)\frac{L_s}{KT} + \frac{R_v T}{e_i(T)D}\right]}. \tag{9.4}$$

Here, as in (7.17), the kinetic effects and ventilation are neglected. These effects are not understood as well for ice crystals as for water droplets. Vapor molecules cannot unite with an ice crystal in any haphazard way, but must join up, molecule-by-molecule, in such a manner that the crystal pattern is maintained. Consequently it may be incorrect to identify ϱ_{vr} with the equilibrium vapor density of ice; and in fact ϱ_{vr} may not be the same over all points of the crystal surface. Because of these effects the rate of growth of an ice crystal will tend to be slower than given by (9.4). Experiments by Fukuta (1969) indicated that at temperatures between about 0° and −10°C the growth rates of small crystals are about half as fast as predicted by (9.4). For large crystal sizes, the formula may be a better approximation. However, ventilation effects become important when the crystal is large enough to fall with a significant velocity relative to air. These effects tend to increase the growth rate, and may be included by modifying the two terms in the denominator of (9.4) with aerodynamic correction factors given by Koenig (1971).

Because the initial growth of ice crystals usually occurs in a water cloud, $S_i \approx e_s/e_i$ in (9.4) with a strong temperature dependence as shown in Fig. 9.3.

The denominator of (9.4) depends on temperature and pressure in approximately the manner indicated in Fig. 7.1 for water droplets. Byers

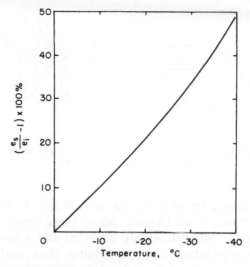

FIG. 9.3. Supersaturation relative to ice in an atmosphere at equilibrium
saturation relative to water.

(1965) combined this temperature dependence with that of S_i, assuming
a water-saturated environment, to determine the dependence of ice
crystal growth rate on temperature and pressure as shown in Fig. 9.4.
These curves indicate that the growth rate varies inversely with pressure,
and that the temperature for maximum growth is about $-15°C$ over a
wide range of pressure.

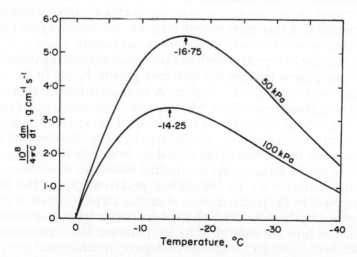

FIG. 9.4. Normalized ice crystal growth rate as a function of temperature.
(Adapted from Byers, 1965.)

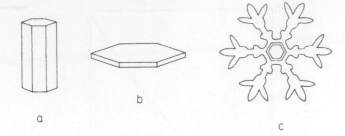

FIG. 9.5. Schematic representation of the main shapes of ice crystals:
(a) column; (b) plate; (c) dendrite.

Ambient conditions determine not only growth rate, but also the form, or habit that a growing crystal takes. All of these forms are basically hexagonal structures, but with widely different axis ratios. Figure 9.5 illustrates the main crystal types, namely column, plate, and dendrite. As a growing crystal moves through a cloud its crystal habit will change according to the changing ambient conditions. Sector stars are formed when plates develop peripheral dendritic structure; capped columns arise when columns develop plates on their ends. The intricate stellar shapes which are often observed are variations on the dendritic form.

Since the pioneering work of Nakaya (1954) several authors have reported on the dependence of crystal habit on environmental conditions. Their work is in good agreement, and is summarized in Fig. 9.6, due to Kobayashi. This figure also indicates the excess vapor density over ice equilibrium in an atmosphere saturated with respect to plane water. This approximates the conditions to be expected in clouds, and shows that the excess vapor density is a maximum at about −15°C, which is also the temperature of maximum growth rate. The preferred crystal types in this growth region are seen to be dendrites and sectors.

The treatment of crystal growth as a process controlled by diffusion of water vapor cannot explain the different growth habits of ice or the transitions from one habit to another. A molecular-kinetic approach is required to explain these effects. Microscopic examination of the surface of a growing crystal shows that it is made up of flat terraces of different heights terminating at ledges and separated by steps. Molecules of water that impinge on the surface of the crystal are bound more strongly at the ledges than on the terraces. As unattached molecules move about over the crystal surface they fix themselves preferentially at the ledges, causing growth by the lateral motion of surface steps. The rate of growth is determined by the rates at which steps are generated and advance, and these rates in turn are controlled by temperature and supersaturation. The main mechanism for the motion of steps is the diffusional movement of molecules across terrace sites up to and into ledge sites. These kinetic

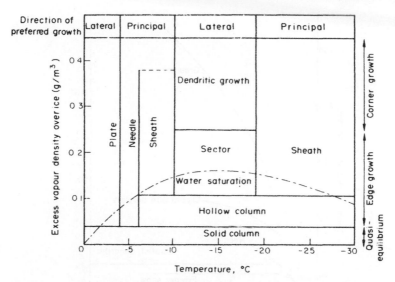

FIG. 9.6. Kobayashi's diagram of crystal habit as function of temperature and excess vapor density over ice saturation. (From Fletcher, 1962.)

effects determine the crystal habit and tend to slow the rate of crystal growth from the value given by the continuum vapor diffusion theory.

Without explaining the shapes of crystals, the classical theory can nevertheless be used for approximate growth calculations if the crystal habit is specified. What is required is an analytical expression for the capacitance of the crystal and empirical relations between the dimensions of the crystal and its mass. Houghton (1985) may be consulted for a compilation of this information.

Further growth by accretion

Strictly speaking, growth by accretion occurs when any large precipitation particle overtakes and captures a smaller one. In common usage, however, *accretion* is sometimes reserved for the capture of supercooled droplets by an ice-phase precipitation particle. If the droplets freeze immediately on contact they form a coating of rime, leading to rimed crystals or graupel. If the freezing is not immediate, denser structures are created, of which hail is an extreme example. *Coalescence* refers to the capture of small cloud droplets by larger cloud drops. It may also be used to describe the capture of cloud droplets by raindrops, although the word accretion is sometimes applied to the water phase when the size difference between the collector drop and the collected droplet is large. *Aggregation* is the clumping together of ice crystals to form snowflakes.

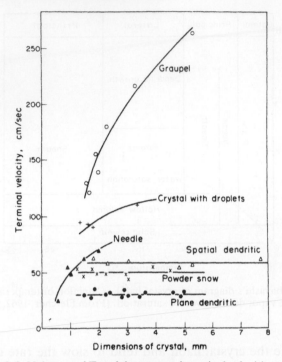

FIG. 9.7. Nakaya and Terada's measured terminal velocities of ice crystals.
(From Fletcher, 1962.)

Important in ice crystal growth by collision and capture is the fall speed
of ice crystals, some of the classical data on which are given in Fig. 9.7.
The fastest falling crystals are graupel particles (which are not really
crystals, but aggregates of frozen droplets). The rimed structures (crystal
with droplets) fall at about 1 m/s, but all the pure crystal types fall slower
than 1 m/s. An empirical formula that provides an approximate fit to the
graupel curve is

$$u = 343D^{0.6} \tag{9.5}$$

with u in cm/s and D, the diameter of the sphere which just circumscribes
the particle, in cm. Snowflakes are not shown on this figure but they are
also found to fall at about 1 m/s as long as they are dry. When melting
begins they become more compact and fall faster. A generally accepted
approximation for snowflake fall speed is due to Langleben (1954),

$$u = kD^n, \tag{9.6}$$

where D is the melted diameter. With D in cm and u the fall speed in
cm/s, Langleben found for dendrites $k \approx 160$ and $n \approx 0.3$, and for
columns and plates $k \approx 234$ and $n \approx 0.3$.

TABLE 9.2. *Values of* a *and* b *in (9.7) for*
$\mathcal{D}$ *in cm and* m *in g*

Crystal type	a	b
Graupel	6.5×10^{-2}	3
Thin hexagonal plate	1.9×10^{-2}	3
Stellar crystal	9.4×10^{-4}	2
Planar dendrite	3.8×10^{-4}	2
Needle	2.9×10^{-5}	1

The mass and size of different forms of ice crystals are usually related by empirical formulas of the form

$$m = a\mathcal{D}^b, \tag{9.7}$$

where $\mathcal{D}$ is the major linear dimension of the crystal. Values of a and b for several typical crystal forms are given in Table 9.2, taken from data in Mason (1971) and Houghton (1985). These references may be consulted for more complete tabulations.

In the process of growth by accretion the question of collection efficiency arises. First there is the aerodynamic problem of collision efficiency; then the question of whether sticking occurs, given a collision. Not much is known about either side of the question. Since ice crystals fall more slowly than water droplets of equal mass, it seems plausible that the collision efficiencies might be higher. However, Pitter and Pruppacher (1974) determined collision efficiencies of simple ice plates theoretically, by calculating trajectories of water droplets relative to the ice crystals, and showed the collection process to be complex, with the perimeter of the crystal a preferred area for collisions. Because freezing is likely to occur on contact with supercooled droplets, the coalescence efficiency might be expected to be unity.

In the process of crystal aggregation, the collection efficiency is less well understood. Indications are that open structures like dendrites are more likely to stick, given a collision, than crystals of other shape, and that sticking in any case is more likely at relatively warm temperatures. Judging from the observed sizes of snowflakes as a function of temperature, it has been inferred that significant aggregation is possible only at temperatures warmer than $-10°C$.

Bearing the uncertainties in mind, we can set up equations to describe growth by these processes. For accretional growth, leading to graupel, an approximation analogous to (8.15) may be employed,

$$\frac{dm}{dt} = \bar{E}M\pi R^2 u(R), \tag{9.8}$$

where m = mass of particle, M = cloud liquid water content, R = radius of particle, $u(R)$ = fall speed, and $\bar{E}$ = mean collection efficiency.

Basically the same approach may be used in analyzing the aggregation process. Since snowflakes all fall at about 1 m/s and ice crystals all fall at about 0.4–0.5 m/s, the growth equation for a snowflake is

$$\frac{dm}{dt} = \bar{E}M\pi R^2 \Delta u, \tag{9.9}$$

where Δu is the difference in fall speed of the snowflake and the ice crystals, essentially a constant. Sometimes the population of ice crystals is more conveniently characterized by the number density N than by the density of frozen water M. They are related by

$$M = Nv\varrho,$$

where v is the (average) volume of the crystals and ϱ is their density. If the snowflake is assumed to have the same density, then $m = \varrho V$ where V denotes its volume. The growth equation in terms of volume then becomes

$$\frac{dV}{dt} = B\bar{E}V^{2/3}Nv\Delta u, \tag{9.10}$$

where $B^3 = \frac{9}{16}\pi$. Clearly these equations must be understood as rough approximations to the actual growth processes. According to Fletcher (1962), calculations based on (9.10) have been found to give results in reasonable accord with observations on graupel and snowflakes.

Fundamentally, snowflakes must develop because a few of the crystals, which formed and grew by diffusion, become larger than their neighbors, either by enhanced diffusional growth or by chance collisions with other crystals or supercooled droplets. Having attained this initial advantage, the crystals or small aggregates are in a favorable position to grow by the sweepout process. A complete theory for the development of precipitation in the ice phase should therefore take into account the statistical effects that are incorporated in the coalescence theory of rain (see Chapter 8). Austin and Kraus (1968) formulated a model for snowflake development on the assumption that random collisions lead to a distribution of aggregate sizes such that gravitational effects can become important. Taking initial values of number density between 10^4 and 10^5 crystals per cubic meter and assuming about 100 collisions per second per cubic meter, they found that reasonable distributions of snowflakes resulted in realistic time periods. The results depend rather critically on the collision frequency, and there seems to be no independent means of establishing 100 as a typical figure for this frequency.

The ice crystal process versus coalescence

For precipitation particles of appreciable size to develop in the time available, it is necessary that aggregation or accretion take place in growth of the ice phase, or that coalescence occur in the all-water process. Condensation-diffusion alone cannot explain the formation of particles as large as a milligram in mass (corresponding to a water radius of 0.62 mm, a moderate-sized raindrop) in realistic times. However, this process is more effective for ice crystals than for water droplets, because the vapor in the cloud is often at equilibrium relative to water and hence supersaturated with respect to ice. It is common experience that light precipitation can occur in the form of individual ice crystals, indicating that aggregation or accretion never occurred. It is reasonable to suppose, therefore, that some of the precipitation reaching the surface in the form of drizzle or very light rain might be due to unaggregated crystals that melted before reaching the ground. In warm clouds on the other hand diffusional growth is too slow to produce even drizzle-size drops in reasonable times; coalescence is always required to produce rain from such clouds.

Many cumulus clouds develop initially at temperatures warmer than 0°C, or at least warm enough to make droplet freezing unlikely, and then grow vertically to altitudes considerably higher than the 0°C level, where ice crystal formation is likely to occur. In such clouds, both precipitation mechanisms may occur—initially the coalescence process among droplets, later the ice crystal process as well. Which process dominates in a given situation depends on the temperature at cloud top, the cloud liquid water content, and to some extent the droplet concentration. The coalescence process will tend to predominate in clouds that are relatively warm with high liquid water contents and low droplet concentrations.

An approximate idea of the difference between the rate of precipitation initiation by the ice-crystal process and by the coalescence of drops can be gained by comparing the early growth history of an ice crystal with that of a large cloud drop. Figure 9.8 illustrates the times required for an ice crystal growing by diffusion and a drop growing by coalescence to reach a certain mass. For this comparison, favorable growth conditions were assumed for both precipitation forms. The ice crystal is a stellar dendrite growing in a water-saturated environment at $-15°C$ and 80 kPa from an initial mass of 10^{-8} g. The capacitance of this form is taken to be $2r/\pi$, with r the radius of the circumscribing disk. Mass and radius are related by the empirical formula given by Houghton (1985), $m = 3.8 \times 10^{-3} r^2$, with m in g and r in cm. The water drop starts with a radius of 25 μm and grows by continuous collection in a cloud consisting of droplets all 10 μm in radius, with a liquid water content of 1 g/m^3. Realistic collection efficiencies are assumed.

The crystal grows rapidly by diffusion, surpassing in size the initially more massive drop at 75 s, and continuing for a while to outpace the drop, whose growth is impeded at first by the small collection efficiency. The dashed curves show the fractional rate of increase of mass, $m^{-1}(dm/dt)$, for the drop and the ice crystal. This quantity is initially

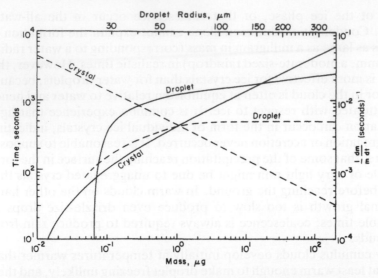

FIG. 9.8. Times required for an ice crystal and a water droplet (solid curves) to grow to the indicated mass. Top scale gives the corresponding drop radius. Dashed curves are for the rates of fractional increase of mass, referred to the scale on the right.

much higher for the crystal, but it decreases as the crystal grows, following nearly exactly a straight line on these coordinates, implying a power-law dependence of dm/dt upon the crystal mass. For the drop, the fractional mass increase starts low and increases monotonically to an asymptotic value, in this example 5×10^{-3} s. For times up to 7 min the ice crystal grows relatively faster than the drop. By this time the drop has finally experienced enough collisions to reach a size where the collection efficiency is no longer small. The drop only reaches the ice crystal in mass at 30 min, when its radius is $160\,\mu m$. This is a drizzle drop, falling at about 1.3 m/s.

Precipitation may be said to be initiated when particles having a mass of approximately $4\,\mu g$ are formed. (For water, this corresponds to the customary threshold radius of 0.1 mm.) The calculation thus shows that in favorable conditions precipitation may be initiated by the ice crystal process in about 10 min or by continuous collection of water in a time about twice as long. In a similar comparison some years ago, Houghton (1950) concluded that in typical midlatitude cumulus clouds precipitation is probably initiated by the ice crystal process because of the slowness of early coalescence growth. Radar observations since that time have indicated that the first (radar detectable) precipitation often appears at cloud levels warmer than 0°C. This is strong evidence that coalescence can initiate precipitation even in clouds that extend vertically to cold temperatures. Moreover, stochastic effects, which are now known to be important in the early stages of coalescence, can reduce the time for precipitation development in the all-water process. It thus appears that, depending on temperature, cloud water concentration, and the droplet sizes, precipitation may be initiated in reasonable times by either process. Subsequent development of the precipitation to significant sizes always requires coagulation—continued coalescence in the water process or accretion and aggregation in the ice process.

Problems

9.1. An ice crystal in the form of a thin hexagonal plate grows by diffusion in an environment saturated with respect to water at a temperature of $-4°C$ and a pressure of 80 kPa. Determine the time required for it to grow to a diameter of 1 mm, starting from a mass of 10^{-8} g. Take the capacity to be that of the circumscribing disk. Assume that the mass and the diameter of the plate are related by $m = 1.9 \times 10^{-2}D^3$, with mass in g and D in cm. Neglect ventilation effects. If diameter and fall speed are related by $u = \varkappa D$, where $\varkappa = 520 \text{ s}^{-1}$, determine the distance the crystal falls during the growth to 1 mm diameter.

9.2. An experiment in freezing nucleation consists of creating a large number of equal-sized droplets from a sample of bulk water and chilling them as a group to subfreezing temperatures. As the temperature is steadily lowered, droplets will begin to freeze: those freezing first contain freezing nuclei active at relatively warm temperatures. With continued chilling, all the droplets will eventually freeze. The experiment is repeated for droplets with a different size, and the temperatures noted where freezing occurs.

For a given droplet size, the median freezing temperature T_m is defined as the temperature at which half the droplets are frozen. This temperature increases with droplet size, because the probability that a droplet contains a freezing nucleus increases with droplet size.

Assume that the number n of freezing nuclei per unit volume of water, which are active at temperatures warmer than or equal to T, increases with the degree of supercooling according to

$$n(T) = N_0 \exp [\lambda(T_0 - T)],$$

where N_0, λ, and T_0 are constants. Solve for T_m as a function of T_0, λ, and droplet diameter. Use the fact that $\exp[-n(T)v]$ is the probability that a droplet of volume v contains no freezing nucleus active at temperature T or warmer.

A set of experimental data shows that T_m increases linearly with the logarithm of droplet diameter, consistent with this freezing model. It shows further that $T_m = -31°C$ for a diameter of $10^2 \mu m$ and $T_m = -17°C$ for a diameter of $10^4 \mu m$. Using this information, solve for λ and T_0.

9.3. The equations that have been given for the diffusional growth of a water droplet or an ice crystal account for heat transfer between the particle and its environment by conduction but not by radiation. For a growing droplet, compare the rates of heat loss by conduction and by radiation, assuming that the droplet and ambient air may be approximated as black body radiators. Show that neglecting radiative heat transfer is a reasonable approximation, and generally a better approximation for droplets than for ice crystals.

10

Rain and Snow

Drop-size distribution

Precipitation may be initiated through either the coalescence process or the ice-crystal process, with coalescence favored in clouds that are relatively warm with high liquid water contents. After precipitation particles are formed they grow chiefly by sweeping out cloud droplets (accretion) or by combining with one another. Depending upon various factors, this continued growth produces raindrops, snowflakes, or hail.

Regardless of how it is initiated, precipitation over much of the world reaches the ground as rain. Its most commonly measured characteristic is the rainfall rate at the surface. A more complete description of the rain is provided by the drop-size distribution function, which expresses the number of drops per unit size interval (usually diameter) per unit volume of space. Such distributions have been measured by a variety of methods in most of the world's climatic regions. Though they are variable in time and space, the distributions usually indicate a rapid decrease in drop concentration with increasing size, at least for diameters exceeding about 1 mm. Also they generally show a systematic variation with rainfall intensity, the relative number of large drops tending to increase with rainfall rate.

Some examples of raindrop spectra are shown in Fig. 10.1. These were obtained with an instrument having a collection area of 50 cm^2 which records the size of individual raindrops by sensing their momentum on impact. For each curve the sample time is indicated as well as the total number of drops counted. A relatively large sample is needed when estimating drop-size distributions to suppress the statistical variability in counts of the rare large drops. In this figure the distributions numbered 1 and 2 were recorded in steady rain; distribution 3 was measured in a thunderstorm.

These examples, and the measurements of many others, indicate that drop-size distributions are of an approximate negative-exponential form, especially in rain that is fairly steady. Marshall and Palmer (1948) first suggested this approximation on the basis of a summer's observations in Ottawa, Canada. Figure 10.2 compares drop spectra at three values of rainfall rate with the best-fit exponential approximations, which are straight lines on the semilogarithmic coordinates. Thus the drop-size distributions, except for very small sizes, may be approximated as

$$N(D) = N_0 e^{-\Lambda D}, \tag{10.1}$$

where $N(D)dD$ is the number of drops per unit volume with diameters between D and $D + dD$. Marshall and Palmer found that the slope factor Λ depends only on rainfall rate and is given by

$$\Lambda(R) = 41R^{-0.21}, \tag{10.2}$$

where Λ has units of cm^{-1} and R is measured in mm/h. Rather remarkably, they also found that the intercept parameter N_0 is a constant given by

$$N_0 = 0.08 \text{ cm}^{-4}. \tag{10.3}$$

It is obvious from Fig. 10.1 that not all drop-size distributions have the simple exponential form. Yet measurements from many different regions

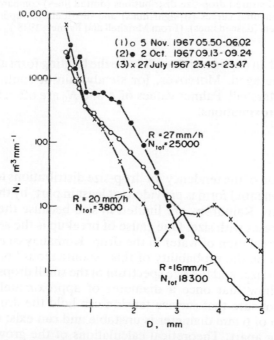

FIG. 10.1. Examples of measured drop-size distributions in rain. Indicated for each curve are the duration of the observation, the total number of drops counted, and the average rainfall rate. Distributions 1 and 2 were recorded during nearly constant rain; distribution 3 was recorded during a thunderstorm. (From Joss *et al.*, 1968.)

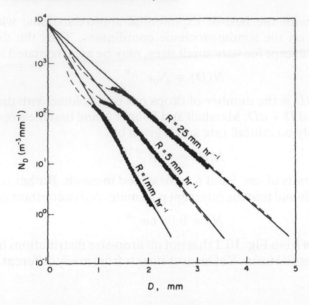

FIG. 10.2. Measured drop-size distributions (dotted lines) compared with best-fit exponential curves (straight lines) and distributions reported by others (dashed lines). (From Marshall and Palmer, 1948.)

have shown that an exponential tends to be the limiting form as individual samples are averaged. Moreover, for steady rain at continental mid-latitudes, the Marshall–Palmer values of Λ and N_0 are often found to be reasonable approximations.

Drop breakup

An explanation of the tendency for drop-size distributions to approach a negative-exponential form is provided, at least in part, by the phenomenon of breakup. Raindrops are limited in size because the chance of disruption increases with size. One cause of breakup is the aerodynamically induced circulation of water in the drop. Komabayasi *et al.* (1964) have given data on the probability of this "spontaneous" breakup as a function of drop size, and the size spectrum of the small drops produced. Their results show that once a diameter of approximately 3 mm is reached, it is not sure that surface tension can hold the drop together, and that a drop of 6 mm diameter is unstable and can exist only briefly before breaking apart. Theoretical calculations of the growth of raindrops including spontaneous breakup have yielded size distributions that are flatter than those observed. Recent theoretical and observational evidence has indicated that water drops may actually reach diameters as large as 10 mm before spontaneous breakup occurs, and it is now thought that this mode of disintegration may be unimportant in the development of raindrop spectra in the atmosphere.

Another cause of breakup is collisions between drops. In a laboratory study of water drops of diameter 0.3 to 1.5 mm colliding at an angle with relative velocities ranging from 0.3 to 3 m/s, Brazier-Smith *et al.* (1972) found that permanent coalescence becomes less likely for increasing values of drop size, relative velocity, and impact parameter. Collisions at grazing incidence produce a spinning, elongated drop which may quickly fly apart, resulting in the formation of satellite drops. Disruption occurs when the rotational kinetic energy of the coupled drops exceeds the surface energy required to produce separate drops. From a comparison of these energies, Brazier-Smith *et al.* obtained the following expression for the coalescence efficiency:

$$\varepsilon = \frac{12\sigma f(R/r)}{5r\varrho U^2} \tag{10.4}$$

where σ is the surface tension of the drops, ϱ is the density, U the relative velocity, and $f(R/r)$ a dimensionless factor given by

$$f(R/r) = f(\gamma) = \frac{[1 + \gamma^2 - (1 + \gamma^3)^{2/3}][1 + \gamma^3]^{11/3}}{\gamma^6(1 + \gamma)^2}. \tag{10.5}$$

For the range of drop sizes tested good agreement was found between (10.4) and the laboratory observations. It should be noted that ε defined by this equation can take on values greater than unity for certain combinations of R, r, and U. Evidently ε may be interpreted as the coalescence efficiency only for values less than or equal to unity.

For drops freely falling in the atmosphere the relative velocity U is the difference in terminal fall speed of drops of radius R and r. The Gunn and Kinzer data of Table 8.2 were used to determine U for all radius pairs. Then (10.4) was evaluated as a function of R and r, and is plotted in Fig. 10.3. This figure shows that the coalescence efficiency is unity for $R < 0.4$ mm or $r < 0.2$ mm. Values of ε less than 0.2 exist for 1 mm $< R < 2.5$ mm. Minima of ε fall approximately on the line $r = 0.6R$. The efficiency is unity for all values of r such that $r < 0.2R$ or $r > 0.9R$. Comparing Figs. 10.3 and 8.3 indicates that the collision efficiency is essentially unity in the region where the coalescence efficiency is less than unity, and conversely. Therefore the *collection* efficiency, to good approximation, equals either the collision efficiency or the coalescence efficiency, whichever is the smaller.

Brazier-Smith *et al.* observed that most of the collisions that did not lead to permanent coalescence resulted in the production of small satellite drops, ranging in number from 1 to 10 and with sizes from 20 to 220 μm radius. In a theoretical study of raindrop interactions and rainfall rates within clouds, the same authors (Brazier-Smith *et al.*, 1973)

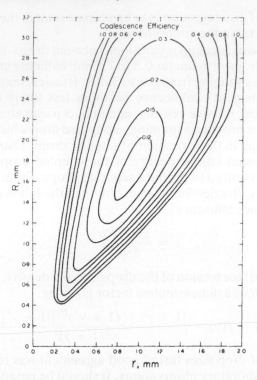

FIG. 10.3. Coalescence efficiency from the theory of Brazier-Smith *et al.*, as a function of drop radii. The value of ε is the fraction of collisions that result in permanent coalescence. All collisions lead to coalescence outside the contour $\varepsilon = 1$.

explained that breakup may be approximated reasonably well by assuming that every collision-induced disruption produces three satellite drops of equal size, each with a volume given by $0.04 V_1 V_2/(V_1 + V_2)$ where V_1 and V_2 denote the volumes of the parent drops. Because of the limited range of radius pairs over which collision-breakup can occur, this formula usually predicts satellite drops of about $100\,\mu m$ radius.

In a theoretical treatment of drop-spectrum evolution, Young (1975) included the effects of collision-induced breakup along with growth by condensation and coalescence. Starting with an assumed activity spectrum of condensation nuclei, he calculated the drop-size distribution as a function of time in a volume of cloudy air ascending at constant velocity. The breakup process was modeled according to the findings of Brazier-Smith *et al.* Thus, after drops grow beyond 0.4 mm some fraction of the collisions cause fragmentation and the production of satellite drops. Figure 10.4 shows the raindrop spectrum after 30 min of ascent at 3 m/s in a cloud with maritime characteristics. The spectrum can be adequately

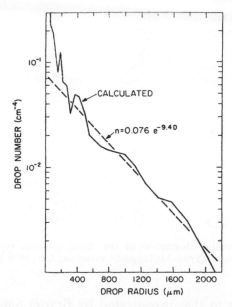

FIG. 10.4. Theoretical drop-size distribution produced after 30 min in a model of droplet growth that includes the effects of condensation, coalescence, and collision breakup. (Adapted from Young, 1975.)

approximated by an exponential of the form (10.1). Although Young pointed out that the exponent Λ in the best-fit curve differs from that predicted by (10.2) for the calculated rainfall rate, the results convincingly demonstrate the importance of collision breakup in establishing the limiting exponential form. The discrepancy in Λ should not be too disturbing in any case, because the Marshall–Palmer relation (10.2) has not been found to fit maritime rain.

The formulation of the breakup process by Brazier-Smith *et al.* has come under question because (1) their experiments were for drops colliding at an angle rather than head-on and (2) the effect of the air was neglected in the theoretical development. Low and List (1982) argued that experiments should be conducted so that the force of gravity, the drop velocities, and the drag force of the air are vertical and parallel, and that the drops should have the correct fall velocity relative to one another and to the air at the time of the collision. They carried out experiments satisfying these requirements for drops with diameters ranging from 0.0395 to 0.46 cm, finding that three common breakup types could be distinguished, as illustrated in Fig. 10.5. The neck or filament breakup is caused by glancing collisions. The identities of the colliding drops are generally preserved and new satellite droplets are created by the disintegration of the connecting neck as the parent drops separate. This mode of

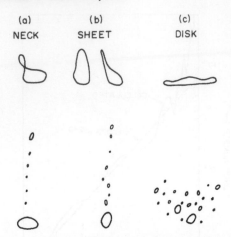

FIG. 10.5. Schematic illustration of the three common types of drop breakup. (From McTaggart-Cowan and List, 1975.)

breakup is similar to that investigated by Brazier-Smith *et al.* Sheet breakup occurs when the drops collide in such a way that one side of the large drop is torn off. The bulk of the large drop then rotates about the point of impact, issuing a sheet of water that breaks apart into satellites. The identity of the large drop is preserved but that of the small drop is lost in this kind of collision. The third type of breakup occurs when the point of impact is close to the center of the bigger drop. Coalescence occurs temporarily but a disk spreads out from the center and disintegrates into a large number of medium-sized drops. The identities of both drops are lost in disk breakup.

In addition to the point of impact on the larger drop, another parameter that was found to determine whether breakup occurs and the subsequent mode of breakup is the collision kinetic energy of the drop pair relative to its center of mass, defined as

$$\text{CKE} = \left(\frac{2}{3}\pi\varrho_L\right)\frac{R^3}{1+\gamma^3}U^2, \tag{10.6}$$

where U is the relative velocity and $\gamma = R/r$. For high values of CKE, all three modes of breakup can occur depending on the point of impact. For smaller CKE and for small values of γ, filament breakup occurs regardless of the point of contact. For very small CKE, coalescence occurs if the total energy E_T, which equals the sum of CKE and the excess surface energy when two drops coalesce to form a bigger drop, can be dissipated through the oscillation and deformation of the drop and the internal circulation of water in the drop. Low and List found experimentally that

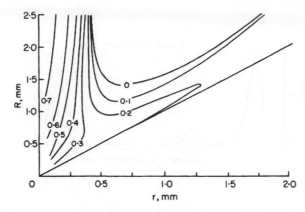

FIG. 10.6. Empirical coalescence efficiencies of Low and List. (Adapted from Low and List, 1982.)

the fraction of collisions leading to coalescence without breakup (that is, the coalescence efficiency) is given by the empirical relation

$$\varepsilon = a\left(\frac{\gamma}{\gamma + 1}\right)^2 \exp\left[-\frac{b\sigma E_T^2}{S_C}\right], \quad \text{for } E_T < 5\ \mu\text{J}, \quad (10.7)$$

$$= 0, \text{ otherwise,}$$

where $a = 0.778$, $b = 2.61 \times 10^6\ \text{J}^{-2}\ \text{m}^2$, σ is the surface tension, and E_T is given by

$$E_T = \text{CKE} + 4\pi\sigma(r^2 + R^2) - S_C, \quad (10.8)$$

with

$$S_C = 4\pi\sigma(r^3 + R^3)^{2/3}.$$

Figure 10.6 shows the coalescence efficiency given by (10.7). The minimum efficiencies lie approximately along the line $R = 2r$, which also delineates the region where CKE is close to a maximum. For a fixed R, the coalescence efficiency generally decreases with increasing r in a manner similar to the results of Brazier-Smith *et al.* However, when r exceeds about 0.5 mm the Low and List value of ε remains small.

Low and List measured the size distributions of the satellite fragments and fitted them with normal and lognormal distributions. Figure 10.7 gives an example of the results for the three breakup modes, for the collision of drops with diameters of 0.46 and 0.18 cm. For filament breakup, besides the two peaks centered at the original sizes, a third peak appears at smaller diameters. Sheet breakup is effective in eliminating the smaller drop size from the distribution. Disk breakup is important in limiting the growth of the largest drops.

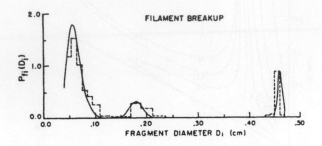

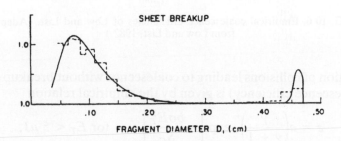

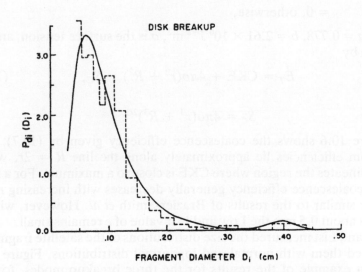

FIG. 10.7. Examples of size distributions of fragments produced by the types of breakup in Fig. 10.5. (Adapted from Low and List, 1982.)

By incorporating the frequency of breakup and the size distribution of fragments deduced from these experiments, Valdez and Young (1985), Brown (1986), and List *et al.* (1987) calculated the drop-size distribution as a function of time as it evolves from an initial Marshall–Palmer distribution. They found that an equilibrium distribution is eventually established in which, for a given amount of rainwater, the shape of the distribution stabilizes. Large drops grow by consuming smaller ones but they eventually break apart, replenishing the numbers of small drops. Figure 10.8 is an example of the results. The dashed lines indicate the initial distributions, corresponding to rainwater contents of 1.766, 3.471, and 5.572 g/m^3 and rainfall rates of 41.0, 81.5, and 134.5 mm/h. These high values of rain amount and rain rate were used to give a high rate of collisions and breakup, minimizing the time required to reach equilibrium. It is evident that the equilibrium spectra, shown by solid lines, deviate significantly from the Marshall–Palmer form. The distributions are essentially parallel to one another and trimodal, with peaks at radii of about 0.12, 0.43, and 1.0 mm. Valdez and Young doubted that these distributions could ever be observed under atmospheric conditions because it took almost an hour to reach equilibrium for a rainwater

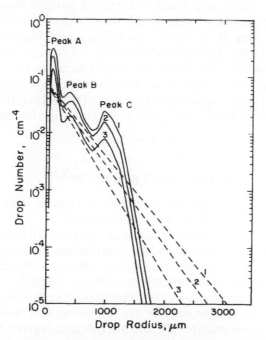

FIG. 10.8. Equilibrium raindrop size distributions for three rainfall rates, and the Marshall–Palmer distributions corresponding to the same three values of liquid water content. (Adapted from Valdez and Young, 1985.)

content of 1.77 g/m^3. Still longer times would be required for smaller and more typical rainwater contents. List *et al.*, on the other hand, presented evidence that the major features of the distribution, especially the location of the peaks, can be established relatively quickly and are present in the computed spectrum 5–10 min after collisions begin. Steiner and Waldvogel (1987) analyzed many measured raindrop spectra and found, rather remarkably, that one or more spectral peaks were often present at four distinct radii: 0.35, 0.5, 0.95, and 1.6 mm. The observed peak at 0.95 mm agrees closely with the theoretical prediction of an equilibrium peak at 1.0 mm by Valdez and Young and by others. The other observed peaks do not correspond as closely to the predictions, although the agreement may be expected to improve as our understanding of the coalescence and breakup process is further refined.

Distribution of snowflakes with size

Snowflakes rather than individual ice crystals account for most of the precipitation reaching the ground as snow. As the snowflakes are irregular aggregates of crystals or smaller snowflakes, there is no easy way to measure their linear dimensions. Consequently, data on snowflake sizes are usually expressed in terms of particle mass or, equivalently, the diameter of the water drop formed when the snowflake melts.

Size distributions of aggregate snowflakes were measured by Gunn and Marshall (1958). They plotted the data on semilogarithmic coordinates, as for raindrops, obtaining the results shown in Fig. 10.9.

Once again the data points for a given rate of precipitation can be fitted reasonably well by an exponential function of the form (10.1). For snow, however, the parameters are related to precipitation intensity by

$$\Lambda \ (\text{cm}^{-1}) = 25.5 R^{-0.48} \tag{10.9}$$

and

$$N_0 \ (\text{cm}^{-4}) = 3.8 \times 10^{-2} R^{-0.87}. \tag{10.10}$$

In these equations the precipitation rate R (mm/h) is in terms of the water-equivalent depth of the accumulated snow.

In theoretical studies of precipitation development it is often necessary to compute moments of the drop-size distribution. For example, the flux of precipitation through a horizontal area, the mass of precipitation per unit volume, and the radar reflectivity of the precipitation are all simply related to certain moments of $N(D)$. This makes the exponential approximation especially convenient to use in theoretical work, since the moments are known analytically: the nth moment is given by

$$\int_0^\infty D^n N(D) dD = N_0 \frac{\Gamma(n+1)}{\Lambda^{n+1}}, \tag{10.11}$$

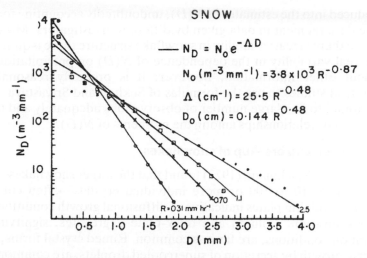

FIG. 10.9. Size distributions of snowflakes in terms of diameters of drops
produced by melting the snowflakes. (From Gunn and Marshall, 1958.)

where Γ denotes the gamma function. (For n an integer $\Gamma(n + 1) = n!$)
This analytical result, which requires an infinite upper limit of integration, is usually a good approximation for real distributions that have a
finite upper limit of diameter. The exponential form of $N(D)$ falls off so
rapidly with D that the unrealistic large particles implied by the infinite
limit make little contribution to the integral.

Reanalyzing the data of Gunn and Marshall and of others, Sekhon and
Srivastava (1970) determined that the negative-exponential function
(10.1) provides an adequate fit for all the observations, but that more
consistent results can be achieved in theoretical work with the moments
of $N(D)$ if the parameters have the values

$$\Lambda \ (\text{cm}^{-1}) = 22.9R^{-0.45} \tag{10.12}$$

and

$$N_0 \ (\text{cm}^{-4}) = 2.5 \times 10^{-2}R^{-0.94}. \tag{10.13}$$

Because of uncertainties in the methods of measuring $N(D)$ for
snowflakes some discrepancies in results are expected. The measurements are obtained from observations of the number of flakes of a given
size that fall on a horizontal surface during a certain exposure time. To
infer from these observations the concentration $N(D)$ of snowflakes in
space, it is necessary to divide the observed distributions by the fall
speeds of the snowflakes in each size interval. These fall speeds are not
uniquely dependent on size, but depend also upon density and possibly
the crystal forms that make up the snowflake. Uncertainty is thus

introduced into the estimates of $N(D)$, undoubtedly accounting for some of the disagreement in data given by different investigators. Moreover, because there is real variability in snowflake structure there is quite likely to be real variability in the dependence of $N(D)$ on precipitation rate. Without other information, however, it is probably reasonable in theoretical studies to use the formulas of Sekhon and Srivastava, which were found to fit a large number of observations adequately and to lead to consistent relationships among the moments of $N(D)$.

Aggregation and breakup of snowflakes

Jiusto and Weickmann (1973) found that the larger snowflakes—those consisting of 10 to 100 or more individual crystals—often consist of dendrites and thin plates indicative of diffusional growth conditions near water saturation. Column and thick-plate aggregates, signifying ice-saturation conditions, are far less common. Rimed crystal forms, which indicate growth by accretion of supercooled droplets, are common when ambient temperatures are relatively mild, and tend to be produced by convective clouds, suggesting the requirement of high liquid water contents and a scarcity of ice nuclei. Jiusto and Weickmann also reported that irregular crystal forms such as fragments, rimed branches, and nonsymmetric segments are sometimes dominant in heavy snowfall.

While it is obvious enough that snowflake size distributions, like those of raindrops, are established by the processes of growth and fracturing, the development of snowflake populations is more difficult to analyze theoretically. The crystal habit is significant in determining the diffusional growth rate, and may also affect the tendency for clumping. Fracturing of snowflakes is likely to be collision-induced, but it may depend on crystal type and temperature. General equations for ice crystal growth by accretion and aggregation have been formulated (see pp. 165–166), but with the many uncertainties about crystal formation and interaction it is unlikely that a comprehensive model of the evolution of size distributions in snow will be developed soon.

Precipitation rates

The precipitation rate or intensity is the flux of precipitation through a horizontal surface. It is measured in terms of the volume flux of water. The SI units are therefore $m^3 \ m^{-2} \ s^{-1} = m/s$, but by convention it is usually expressed in mm/h.

The intensity can be written in terms of the size distribution function $N(D)$ as

$$R = \frac{\pi}{6} \int_0^\infty N(D)D^3u(D)dD, \qquad (10.14)$$

where $u(D)$ is the fall velocity of particles of size D. With the convention that D refers to melted diameter and R to the equivalent rainfall rate, (10.14) applies to snow as well as rain.

A measure of the precipitation amount that is independent of the fall speed is the precipitation water content L, defined by

$$L = \frac{\pi}{6} \varrho_L \int_0^\infty N(D)D^3 dD. \tag{10.15}$$

Values of R at the surface can vary from trace amounts up to several hundred mm/h. Rainfall rates in excess of about 25 mm/h are always associated with convective clouds. At most localities precipitation rates in the form of snowfall tend to be at least an order of magnitude less than those in the form of rainfall. For the Montreal area, which may be fairly typical of many midlatitude localities, the main contribution to a year's total rainfall comes from rainfall rates of about 10 mm/h.

Problems

10.1. For a population of raindrops having an exponential distribution,

$$N(D) = N_0 e^{-\Lambda D},$$

it is easy to show that the median diameter, D_m, is equal to $(\ln 2)/\Lambda$. In theoretical work, a quantity sometimes used to characterize a drop population is the median volume diameter, D_0, defined such that the drops with diameters smaller than D_0 account for half the total volume of rainwater. Solve for D_0 as a function of D_m.

10.2. A small graupel particle is caught in the updraft of a developing cumulonimbus cloud. Its initial diameter is 0.1 mm and the updraft is constant at 10 m/s. Determine the time required for the particle to grow to a diameter of 2 mm and the distance through which it ascends during this time. Assume the cloud liquid water content is constant at 2 g/m^3. Neglect growth by sublimation.

The relation between the mass and radius of graupel is approximately

$$m = 0.52r^3 \quad \text{(CGS units)}$$

and the dependence of graupel fall speed on size may be approximated as

$$u(r) = 520r^{0.6} \quad \text{(CGS units)}.$$

Assume a collection efficiency of unity.

10.3. The rate of production of cloud water at a given level in a developing convective cloud is proportional to the updraft speed at that level. As rain falls through the level it sweeps out cloud droplets and depletes the cloud water content. Derive the relationship between rainfall rate and updraft speed that corresponds to a balance between the rate of cloud water production and the rate of sweepout. Make the following assumptions:

(1) All vapor excess over the equilibrium saturation ratio immediately condenses to form cloud droplets.
(2) Raindrop size distribution is of the general exponential form

$$N(D) = N_0 \exp(-bD),$$

with N_0 a constant equal to the Marshall–Palmer value 0.08 cm^{-4} and b a parameter that depends on the rainfall rate.
(3) Terminal fall speed of the raindrops is related to drop diameter by

$$u(D) = kD,$$

with $k = 4 \times 10^3$ s^{-1}.

11

Weather Radar

MICROWAVE radar was developed early in World War II to aid in spotting distant ships and airplanes. It was noticed from the first that during disturbed weather conditions a kind of widespread interference often appeared on the radar screen and obscured the military objects of interest. A large body of theoretical and experimental work in the decade of the 1940s showed that this "weather clutter" arose from the scattering of radar waves by precipitation. These early findings have now been refined and elaborated to the point that most of the measurable properties of radar signals—amplitude, phase, polarization, and frequency—can be interpreted in terms of the sizes, shapes, motions, or thermodynamic phase of the precipitation particles. Because of their ability to observe and measure precipitation quickly, accurately, and from great distances, radars have become essential in cloud physics research and in weather observation and forecasting. Many radar systems have been designed and constructed specifically for meteorological applications.

This chapter outlines the essentials of radar meteorology and provides a background for the radar examples in later chapters. The standard text on the subject is that of Battan (1973). Accounts of the early history of radar in meteorology are included in Atlas (1989). Recent developments, with emphasis on Doppler techniques, are presented by Doviak and Zrnic (1984).

Principles of radar

The main components of a radar are the transmitter, antenna, and receiver. The transmitter generates short pulses of energy in the radio-frequency portion of the electromagnetic spectrum. These are focused by the antenna into a narrow beam. They propagate outwards at essentially the speed of light. If the pulses intercept an object with different refractive characteristics from air, a current is induced in the object which perturbs the pulse and causes some of the energy to be scattered. Part of the scattered energy will generally be directed back

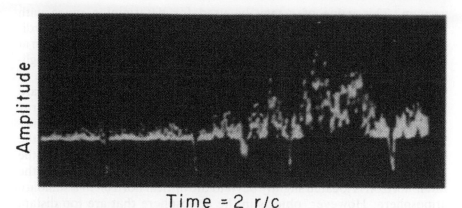

FIG. 11.1. Radar A-scan. A weather target is present in the background of receiver noise.

toward the antenna, and if this backscattered component is sufficiently large it will be detected by the receiver.

The primary function of radar is to measure the range and bearing of backscattering objects or "targets". Ranging is accomplished by a timing circuit that counts time between the transmission of a pulse and the reception of a signal. Direction is determined by noting the antenna azimuth and elevation at the instant the signal is received.

The fundamental radar display is the A-scope, an oscillograph trace of returned signal amplitude versus time after pulse transmission, as shown in Fig. 11.1. Because the energy travels with velocity c, the time interval t between transmission and reception is related to target range by $r = ct/2$. The factor $1/2$ arises because the energy must make a round trip to range r in time t.

Timing begins from the initial transmission of each pulse, which under some conditions can lead to ambiguous range determinations. Suppose that a target is located so far from the transmitter that the return from a particular pulse is not received until after another pulse has been transmitted. In this case an erroneously close range is indicated. For a given radar pulse repetition frequency (PRF) there is a maximum range within which targets will be correctly indicated. Targets beyond this r_{max} that return enough energy to be detected will be displayed ambiguously at a range closer than r_{max}. This maximum unambiguous range is given by $r_{max} = c/2f_r$, where f_r denotes the PRF.

The velocity of propagation c of radar waves in the atmosphere depends on the density and the water vapor content of the air but is always within about 0.03% of 3×10^8 m/s, the velocity of light in a vacuum. This value of c may be used for determining range with acceptable accuracy for all practical purposes. The atmosphere is ordinarily stratified in such a way that c is slightly less near the ground than at higher altitudes. Although the effect on range determination is negligible, the increase of c with altitude causes a downward refraction of waves propagating in directions close to the horizontal. For a standard atmosphere, and over distances normally used in radar detection, the ray paths are approximately circular arcs with a radius of curvature somewhat greater than that of the earth. As a convenient approximation, the rays may be regarded as straight lines in a standard atmosphere if the actual radius of the earth R_E is increased to $(4/3)R_E$. This means that the radar horizon is located slightly beyond the optical horizon in a standard atmosphere. However, objects in the atmosphere that are too distant horizontally are lost in the earth's shadow, invisible to radar as they are to the eye. When the vertical structure of the atmosphere deviates significantly from the standard conditions (for example when strong inversions or elevated moist layers are present), nonstandard or anomalous propagation can occur. Abnormal downward bending of the radar waves ("superrefraction") is a kind of anomalous propagation that occasionally confuses the interpretation of weather echoes.

Some of the more important radar parameters, and their range of values for typical weather radars, are as follows:

1. Peak power (the instantaneous power in a pulse), P_t,

$$10 < P_t < 10^3 \text{ kW.}$$

2. Radio frequency, ν,

$$3 < \nu < 30 \text{ GHz}$$

(corresponding to wavelengths between 1 and 10 cm).

3. PRF, f_r,

$$200 < f_r < 2{,}000 \text{ s}^{-1}.$$

4. Pulse duration, τ,

$$0.1 < \tau < 5 \text{ } \mu\text{s.}$$

An additional radar parameter of importance in meteorological work is the beamwidth, which is determined by the wavelength and the antenna size and shape. Beamwidth is defined with reference to the antenna pattern, which is a plot of the radiated intensity as a function of angular distance from beam axis. Such patterns will generally be different for planes through the axis having different orientations. Many antennas used with weather radars are paraboloidal, however, and the beam patterns are about the same for all planes through the axis. The beamwidth is usually defined as the angular separation between points where the transmitted intensity has fallen to half its maximum value, or 3 dB below the maximum. A typical beamwidth for a meteorological radar is 1 degree.

Antenna patterns are characterized by sidelobes which are in general undesirable but unavoidable. Antenna design is concerned with achieving satisfactory compromises between characteristics of the mainlobe and the sidelobes.

The radar equation

The amount of power returned by a radar target determines whether it will be detectable. The radar range equation expresses the relationship between the returned power and characteristics of the radar and the target. We consider first a target of negligible spatial extent, called a point target.

Suppose the radar transmits a peak power P_t. If this were radiated isotropically, a small area A_t at range r would intercept an amount of power given by

$$P_\sigma = \frac{P_t A_t}{4\pi r^2}.$$

The antenna is used to focus the energy in a narrow beam, increasing the power relative to the isotropic-radiated value. If centered on the beam axis, the small area A_t intercepts an amount of power given by

$$P_\sigma = G\,\frac{P_t A_t}{4\pi r^2},$$

where G is a dimensionless number called the antenna axial gain.

Now, if this area were to scatter the incident radiation isotropically, the power returned to an antenna with aperture area A_e would be

$$P_r = \frac{P_\sigma A_e}{4\pi r^2} = \frac{G P_t A_t A_e}{(4\pi r^2)^2}.$$

Because the gain and the antenna aperture are approximately related by

$$G = \frac{4\pi A_e}{\lambda^2},$$

it follows that

$$P_r = P_t\,\frac{G^2 \lambda^2}{(4\pi)^3 r^4}\,A_t.$$

Most targets do not scatter isotropically, however, and as a convenient artifice the radar backscatter cross-section σ of the target is introduced, and defined by

$$P_r = P_t \frac{G^2 \lambda^2}{(4\pi)^3 r^4} \sigma. \tag{11.1}$$

This is the form of the radar equation for a single target of backscatter cross-section σ. (Note that in general $\sigma \neq A_t$.)

The weather radar equation

Raindrops, snowflakes, and cloud droplets are examples of an important class of radar targets known as distributed targets. Such targets are characterized by the presence of many effective scattering elements that are simultaneously illuminated by a transmitted pulse. The volume containing those particles that are simultaneously illuminated is called the resolution volume of the radar, and is determined by beamwidth and pulse length. For distributed targets whose scattering elements move relative to each other, the power returned from a given range is observed to fluctuate in time. Such fluctuations occur in weather radar signals because the raindrops or snowflakes move relative to one another owing to different fall speeds and wind variations across the resolution volume. The instantaneous power of the fluctuating signal depends upon the arrangement of the scatterers at that time and is not simply related to their backscatter cross-sections. It turns out, however, that a suitably long time average (in practice about 10^{-2} s) of the received power from a given range is given by

$$\bar{P}_r = P_t \frac{G^2 \lambda^2}{(4\pi)^3 r^4} \sum \sigma, \tag{11.2}$$

where $\sum \sigma$ is the sum of the backscatter cross-sections of all the particles within the resolution volume. This "contributing volume" is given approximately by

$$V = \pi \left(\frac{r\theta}{2} \right)^2 \frac{h}{2}, \tag{11.3}$$

where $h = c\tau$ is the pulse length and θ is the beamwidth.

Sometimes (11.2) and (11.3) are combined to give

$$\bar{P}_r = P_t \frac{G^2 \lambda^2}{(4\pi)^3 r^4} \pi \left(\frac{r\theta}{2} \right)^2 \frac{h}{2} \eta, \tag{11.4}$$

where η denotes the radar reflectivity per unit volume.

Both (11.2) and (11.3) (and consequently 11.4) assume that the antenna gain is uniform within its 3-dB limits, which is not true. In (11.2) an average gain, somewhat less than the axial gain, should be employed; also the effective volume could be defined as an integral over the beam pattern instead of simply as the region within 3-dB beam limits. On the assumption of a Gaussian beam pattern, (11.4) in more accurate form becomes

$$\overline{P}_r = P_t \frac{G^2 \lambda^2 \theta^2 h}{1024 \pi^2 \ln 2} \frac{\eta}{r^2}. \qquad (11.5)$$

This differs by a factor $1/(2 \ln 2) = 0.72$ from (11.4).

For a single spherical scatterer that is small compared to the radar wavelength (about 0.1λ is small enough), the backscatter cross-section is related to the sphere radius r_0 by

$$\sigma = 64 \frac{\pi^5}{\lambda^4} |K|^2 r_0^6, \qquad (11.6)$$

where $K = (m^2 - 1)/(m^2 + 2)$ and $m = n - ik$ is the complex index of refraction of the sphere, with n = refractive index and k = absorption coefficient. This is called the Rayleigh scattering law; particles small enough for it to apply are called Rayleigh scatterers. Raindrops and snowflakes may be regarded as Rayleigh scatterers to good approximation at wavelengths of 5 cm and 10 cm, which are common for weather radars. At 3 cm, also a popular wavelength, the approximation is still used but is less accurate. Large hailstones deviate from Rayleigh scattering behavior even at the long wavelength of 10 cm. The refraction term K in (11.6) depends on temperature, wavelength, and the composition of the sphere. For the wavelengths employed in weather radars, and over the meteorological range of temperatures, $|K|^2 \approx 0.93$ for water and 0.21 for ice. Therefore an ice sphere has a radar cross-section only about 2/9, or 6.5 dB less than, that of a water sphere of the same size.

For a collection of spherical raindrops small compared with the wavelength, which are shuffling about, the average received power is

$$\overline{P}_r = P_t \frac{G^2 \lambda^2}{(4\pi)^3 r^4} 64 \frac{\pi^5}{\lambda^4} |K|^2 \sum r_0^6,$$

where Σ again is a summation over the contributing volume. In terms of the drop diameters

$$\overline{P}_r = P_t \frac{G^2 \pi^5}{(4\pi)^3 r^4 \lambda^2} |K|^2 \sum D^6.$$

Thus for spherical scatterers small with respect to the wavelength, the mean power received is determined by radar parameters, range, and by only two factors that depend upon the scatterers: the value of $|K|^2$ and the quantity ΣD^6. Because of the significance of the latter factor a new quantity Z is introduced, defined by

$$Z = \sum_v D^6 = \int_0^\infty N(D)D^6 dD, \tag{11.7}$$

where Σ_v denotes a summation over *unit* volume and $N(D)dD$ is the number of scatterers per unit volume with diameters in dD. For raindrops $N(D)$ is the drop-size distribution. For snowflakes $N(D)$ is the distribution of melted diameters. (If this convention were not adopted for snow, the density of the snow would have to appear as a correction to $|K|^2$.)

In terms of Z, the radar equation, including the small correction for a Gaussian beam pattern, becomes

$$\bar{P}_r = \frac{\pi^3 c}{1024 \ln 2} \left[\frac{P_t \tau G^2 \theta^2}{\lambda^2} \right] \left[|K|^2 \frac{Z}{r^2} \right]. \tag{11.8}$$

$$\underset{\text{RADAR}}{} \quad \underset{\text{TARGET}}{}$$

This is the most useful form of the radar equation, with the radar parameters shown separate from the target parameters.

Following (11.8), the received power may be related to the reflectivity factor Z by

$$10 \log \bar{P}_r = 10 \log Z - 20 \log r + C, \tag{11.9}$$

where C is a constant—something like a sensitivity factor—determined by radar parameters and the dielectric character of the target. In this logarithmic form of the equation the power in decibels is related to the reflectivity factor as measured on a decibel scale. Usual conventions are that $\bar{P}_r$ is measured in milliwatts, with the quantity $10 \log \bar{P}_r$ called the power in dBm (decibels relative to a milliwatt), and Z is measured in mm^6/m^3 with the quantity $10 \log Z$ called the reflectivity factor in dBz. The logarithmic version of the equation is useful because of the wide ranges over which $\bar{P}_r$ and Z vary.

Some energy is lost from a radar beam by absorption and scattering by atmospheric constituents. Clouds and precipitation absorb and scatter a fraction of the microwave energy incident upon them; the gaseous constituents oxygen and water vapor absorb weakly over the microwave spectrum. These effects depend on the radar wavelength, and are generally more severe for the shorter wavelengths. For quantitative measurements of reflectivity, the attenuation due to scattering and absorption should be taken into account. For this reason, the relatively long wavelengths of 5 and 10 cm are favored, for which the attenuation effects are usually small.

Relation of Z to precipitation rate

From its defining equation (11.7), Z depends on the drop-size distribution and is very sensitive to the large-drop component of the distri-

TABLE 11.1. *Reflectivity as a Function of Rainfall Rate*

R (mm/h)	0.1	1	10	100
Z (mm^6/m^3)	5	200	7950	316,000
dBz	7	23	39	55

bution. For a Marshall–Palmer distribution of raindrops extending from zero diameter to infinity, the reflectivity factor is given by

$$Z = N_0 \frac{6!}{\Lambda^7} = N_0 \frac{6!}{(41)^7} R^{1.47}.$$

This is in fair agreement with empirical data on Z and R for rain, which show that generally

$$Z = 200R^{1.6} \tag{11.10}$$

to a reasonable approximation. Table 11.1 gives examples of Z values for several rainfall rates based on (11.10).

A fundamental radar limitation is the noise level of the receiver. Without special provision, the returned signal is not detectable unless it is stronger than the noise. Well-designed receivers have noise levels of about -105 to -110 dBm. For typical weather radars the values of the sensitivity factor C are such that the minimum detectable rainfall rate for a range of about 10 miles is in the order of 0.1 mm/h, corresponding to drizzle. Consequently, weather radars usually detect rain but not cloud. For cloud studies, special radars with wavelengths of about a centimeter or shorter are employed.

There is more variability among the Z–R relations for snow than for rain, but an approximate relation that is generally accepted is

$$Z = 2000R^2, \tag{11.11}$$

where, as in (11.10), R denotes precipitation rate in mm of water per hour.

Empirical relations have also been determined between the reflectivity factor and the precipitation content, L. For rain, an often used relation is

$$Z = 2.4 \times 10^4 L^{1.82} \tag{11.12}$$

and for snow,

$$Z = 3.8 \times 10^4 L^{2.2}, \tag{11.13}$$

where L is in g/m^3 and Z in mm^6/m^3.

Radar displays and special techniques

The most common display is the PPI (Plan Position Indicator) which maps the received signals or "echoes" on polar coordinates in plan view.

With elevation angle fixed, the antenna scans 360° in azimuth with the beam sweeping across a conical surface in space. At every azimuth the voltage output of the receiver as a function of range is used to intensity-modulate a tube with polar coordinates. The distribution of precipitation in plan view is thereby produced, and a time sequence of PPIs indicates the development and motion of precipitation areas. One full azimuth scan requires in the order of 10 s and photographic records are usually kept at the rate of one frame per revolution.

Without careful calibration and maintenance procedures, PPI records do little more than show where and when precipitation is occurring and indicate roughly where the rain is relatively intense (bright echoes). This information in itself is useful in synoptic meteorology and in cloud physics investigations, but is of limited value in quantitative precipitation studies. For this work, it is desirable to know the actual distribution of Z within the echoes. In principle this information can be obtained from PPI film records by densitometry or some equivalent form of exposure analysis. In practice it is extremely difficult to maintain the required overall system calibration—from radar receiver to film processing—for this approach to be accurate.

Similar to the PPI, the RHI (Range Height Indicator) is a display which is generated when the antenna scans in elevation with azimuth fixed. While the PPI emphasizes horizontal echo structure, the RHI shows the details of vertical structure.

Three-dimensional echo coverage can be achieved by a programmed antenna scan, in which azimuth and elevation are systematically varied to survey all or most of space around the radar site. Analog techniques have been used to combine the data from a spiral scan automatically, to produce constant-altitude PPI (CAPPI) maps at several levels above the ground. A three-dimensional scan requires more time than the simple PPI or RHI—usually about 5 min.

Rapid developments in computer technology since the 1970s have made it possible to process radar data digitally, avoiding many of the uncertainties in analog recording. Typically, the returned signals are averaged over a few successive pulses, digitized, and stored as a matrix of values of $\bar{P}_r$ in range-azimuth-elevation space, the natural coordinates of the radar. The resolution of this discrete description of a continuous field is typically one degree in azimuth and elevation and a few hundred meters in range. The three-dimensional structure of the precipitation observed by a radar operating with a 200-km range might thus consist of the average power in as many as 3×10^6 "bins". At approximately every 5 min, the data are updated as a new scanning cycle is completed. Microprocessors are used to convert the measurements to reflectivity factor Z or rainfall rate R, and to generate maps or other derived fields from the data.

In modern radar sets, it is not uncommon for the transmitter to be coherent. By this it is meant that the frequency of the transmitted signal is constant and that each pulse bears the same phase relation to its predecessor. This is the kind of signal produced by pulse-modulating a free-running stable oscillator; it is a characteristic of radars having klystron transmitters. Using such equipment, and making special provisions in the receiver, it is possible to take Doppler velocity measurements. In effect, the frequency content of the returned signal is compared with that of the transmitted signal, and frequency shifts are interpreted as arising from the Doppler effect. Thus a frequency shift $\Delta\nu$ corresponds to a velocity $\vec{V}$ according to

$$\Delta\nu = \frac{2}{\lambda} \vec{V} \cdot \hat{r},$$

where $\hat{r}$ denotes a unit vector in the radar-pointing direction, and $\vec{V} \cdot \hat{r} = v_r$, the radial velocity.

Meteorological targets induce a spectrum of Doppler shifts because their scattering elements generally do not all move with the same velocity. Most of the work in Doppler radar studies is concerned with meteorological interpretations of the Doppler spectrum. When the beam is pointed vertically, the Doppler spectrum contains information about vertical air motions and precipitation fall speeds. For horizontal viewing, Doppler velocities are interpreted as arising from horizontal air motions. The scattering particles move with the wind to a close approximation, though the echo systems frequently do not move exactly with the wind because of precipitation development or dissipation in preferred regions of the echo.

To measure target reflectivity, the amplitude of the returned signal is compared with that of the transmitted signal. In Doppler velocity measurements, the frequencies of returned and transmitted signals are compared. It is also possible to derive information about the target by comparing the polarization of the received and transmitted waves. Non-symmetrical scattering objects induce an amount of cross-polarization depending in a complex way on their shapes, sizes (compared with a wavelength), and dielectric properties. For precipitation, a theory has been developed which relates the cross polarization to the axial ratio of the particles, which are approximated as ellipsoids small compared to the wavelength. There is increasing evidence that polarization techniques provide a method of distinguishing between rain and other precipitation forms (Rogers, 1984).

Problems

11.1. Consider the following raindrop population:

    (i) Drop-size distribution given by

$$N(D) = N_0 \exp(-bD), \quad 0 \leq D \leq \infty,$$

where $N_0 = 0.08 \text{ cm}^{-4}$, a constant, and b is a parameter that depends on rainfall rate.

(ii) Fall speed approximated by

$$u(D) = kD,$$

where $k = 4 \times 10^3 \text{ s}^{-1}$.

For this model, calculate the relationship between radar reflectivity factor Z, in mm^6/m^3, and rainfall rate R, in mm/h.

11.2. For the exponential model of problem 11.1, solve for the relation between Z and R if the fall speed is approximated by

(a) $u(D) = KD^{1/2}$, where $K = 1420 \text{ cm}^{1/2} \text{ s}^{-1}$;
(b) $u(D) = A - B\exp(-CD)$, where $A = 965 \text{ cm/s}$; $B = 1030 \text{ cm/s}$; and $C = 6 \text{ cm}^{-1}$.

Compare the Z/R relations for these two models with the Marshall–Palmer formula, $Z = 200R^{1.6}$.

11.3. The Doppler spectrum S of a weather target indicates the way the returned power is distributed over radial velocity, v_r. For vertical viewing of raindrops falling through still air, the spectrum may be written

$$S(v_r) = S(u) = \frac{\bar{P}_r}{Z} D^6 N(D) \frac{dD}{du},$$

where $u(D)$ is the terminal fall velocity of drops of diameter D, and Rayleigh scattering is assumed. The convention used here is that v_r is positive towards the radar (downwards in this case) and $S(u)$ is normalized in the sense that

$$\int_0^\infty S(u)du = \bar{P}_r.$$

The mean Doppler velocity is defined by

$$\langle u \rangle = \frac{1}{\bar{P}_r} \int_0^\infty uS(u)du.$$

For the exponential drop-size distribution of problem 11.1 and a fall speed given by $u(D) = KD^{1/2}$ as in problem 11.2, show that $\langle u \rangle$ and Z are related by

$$\langle u \rangle = 3.8Z^{1/14},$$

where $\langle u \rangle$ is in m/s and Z is in mm^6/m^3.

11.4. At a time early in the development of a cumulus congestus cloud, the radar reflectivity equals -30 dBz and the droplet spectrum has a Gaussian shape, centered at radius $8 \mu\text{m}$ and with a dispersion of $\sigma/\bar{r}$ of 0.15. Assume that the droplets grow only by condensation, neglecting the solution and curvature terms in the growth equation. Solve for the reflectivity factor (in dBz) at 5, 10, and 15 min later, assuming a constant supersaturation of 0.5% at a temperature of 0°C and a pressure of 70 kPa.

11.5. A population of raindrops consists of N_0 drops per unit volume, all of diameter D_0. The radar reflectivity factor, rainfall rate, and liquid water content are, respectively, Z_0, R_0, and L_0. If the same amount of water is spread over N equal-sized drops per unit volume, with $N \neq N_0$, solve for the new reflectivity factor Z in terms of N, N_0, and Z_0. If, further, the drops are in the size range where the linear fall speed law of problem 11.1 applies, show that $Z \propto R^3$.

11.6. The following simple model is found to be satisfactory for explaining certain radar observations of Hawaiian orographic rain:

(i) Rainwater content L is constant throughout the cloud.

(ii) At any altitude all raindrops are the same size.

(iii) The size of the drops increases by coalescence as they fall. This causes a reduction in their concentration given by

$$\frac{dN}{dz} = \alpha N,$$

where N is the number density of raindrops at altitude z and α is a constant.

Show that in a steady state the reflectivity decreases with altitude at a rate, in units of decibels per unit distance, equal to 4.34α.

11.7. Waldvogel (1974) gave the following data as an example of a rain sample. Entered here are the number of raindrops in the indicated size intervals that fell in a 1-min period on an instrument with a horizontal sampling area of 50 cm^2.

Diameter interval (mm)	Number of drops
0.3–0.4	45
0.4–0.5	39
0.5–0.6	55
0.6–0.7	75
0.7–0.8	84
0.8–1.0	195
1.0–1.2	129
1.2–1.4	53
1.4–1.6	13
1.6–1.8	4
1.8–2.1	3
2.1–2.4	1

For this sample, calculate the rainfall rate in mm/h and the reflectivity factor in mm^6/m^3. For the dependence of fall velocity on size, use the data of Gunn and Kinzer in Table 8.1.

12

Precipitation Processes

THE spatial extent, intensity, and lifetime of a precipitation system are mainly controlled by vertical air motions. To a large extent, these air motions are the manifestation of the different types of atmospheric instabilities described in Chapter 3. The characteristic time and length scales associated with the instabilities depend upon the following atmospheric properties:

D, the depth of the unstable layer ($D \lesssim 10$ km)
N, the Brunt–Väisälä frequency ($N \sim 10^{-2}\,\text{s}^{-1}$)
f, the Coriolis parameter ($f \sim 10^{-4}\,\text{s}^{-1}$ at midlatitudes)
$\beta = \partial f/\partial y$, the meridional gradient of f ($\beta \sim 10^{-11}\,\text{m}^{-1}\,\text{s}^{-1}$)
$\partial u/\partial z$, the wind shear ($\partial u/\partial z \sim 2 \times 10^{-3}\,\text{s}^{-1}$)

The methods of scale analysis (e.g., Emanuel, 1986) and linear stability theory (e.g., Emanuel, 1979), details of which are beyond the scope of this book, lead to the estimates in Table 12.1 of the characteristic length and time scales of motions associated with the different kinds of instability.

It is clear that baroclinic instability accounts for precipitation systems on the synoptic scale. The vertical motion is widespread and in the order of centimeters per second. Conditional instability produces cumulus convection with localized updrafts of several meters per second. Symmetric instability can account for mesoscale systems with updraft speeds between those of the convective and synoptic scale systems.

TABLE 12.1. *Length and Time Scales of Major Instabilities*

Type	Horizontal scale	Time scale
Conditional instability	$D \leq 10$ km	$\dfrac{1}{N} \sim 8$ min
Symmetric instability	$\dfrac{(\partial u/\partial z)D}{f} \leq 200$ km	$\dfrac{1}{f} \sim 3$ h
Baroclinic instability	$\dfrac{f^2(\partial u/\partial z)}{N^2\beta} \sim 2000$ km	$\dfrac{2\pi N}{(\partial u/\partial z)f} \sim 3$ days

The close relation between the scale and intensity of vertical motions and the character of the precipitation they produce leads to the classification of precipitation as one of two types, depending on the dominant mechanism responsible for the vertical motion.

1. Widespread, stratiform, continuous precipitation associated with large scale ascent produced by frontal or topographic lifting or large scale horizontal convergence.

2. Localized, convective, showery precipitation associated with cumulus-scale convection in unstable air.

This is a useful classification, although the distinction between stratiform and convective precipitation is not always sharp. Widespread precipitation, when observed either by radar or raingauge, invariably shows fine-scale structure with the most intense precipitation confined to elements with a size of only several kilometers. Precipitation of convective origin can extend over a large area and produce a pattern similar to that of continuous precipitation. Nevertheless, it is usually possible to describe a pattern as either markedly nonuniform (hence convective), with locally intense regions ranging in size from 1 to 10 km and separated from one another by areas free of precipitation, or rather uniform (hence stratiform) with less pronounced small scale structure and a wider overall extent. Moreover, the pattern of stratiform precipitation evolves relatively slowly in time, and that of convective precipitation changes rapidly.

Stratiform rain is produced in nimbostratus clouds, although dissipating cumulus clouds and orographic clouds may contain rain with stratiform structure. Most snow originates in nimbostratus clouds, but snow flurries and graupel showers can be produced in convective clouds.

Stratiform precipitation

Figure 12.1 is an example of a radar pattern of stratiform precipitation. This is a slant-PPI at the rather high elevation angle of 15 degrees. Four or five shades of gray are discernable, each corresponding to an interval of 10 dBz in reflectivity factor, as indicated along the calibration scale at the bottom. The circular symmetry of the pattern implies a reflectivity field that is horizontally stratified. At ranges beyond 5 km, corresponding to an altitude of 1.3 km above the radar site, the precipitation is in the form of snow. At closer ranges the temperature is warmer than 0°C and the snow has melted and turned to rain.

More details of the vertical structure of this precipitation are shown in Fig. 12.2, an RHI display at an azimuth angle of 30 degrees. The narrow band of high reflectivity just above 1 km is the radar "bright band", which indicates the melting layer. The intersection of this band with the radar beam at an elevation angle of 15 degrees accounts for the bright ring in

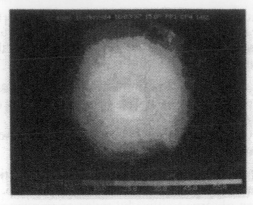

FIG. 12.1. Plan view of stratiform precipitation at an elevation angle of 15 degrees. Radar operated by National Center for Atmospheric Research, and located near Boulder, Colorado. In this computer-generated display, reflectivity is shown in several shades of gray, each corresponding to an interval of 10 dBz as indicated on the calibration strip. Range rings are at 10-km intervals. (Courtesy of NCAR.)

Fig. 12.1. Figure 12.2 shows that the reflectivity pattern in the rain is less uniform horizontally than that in the snow. This tendency is often observed, which means that the rain at the ground is variable even though the snow aloft appears uniform.

The bright band is one of the distinctive features of stratiform rain patterns; other examples are shown in Fig. 12.3. The vertical profile in this figure is for a typical melting layer. It shows that the layer of strengthened reflectivity is several hundred meters thick and that the reflectivity maximum can exceed by 10 dBz the reflectivities in the rain

FIG. 12.2. Vertical cross section at an azimuth angle of 30 degrees through the precipitation in Fig. 12.1. The maximum reflectivities in the melting layer are near 45 dBz. (Courtesy of NCAR.)

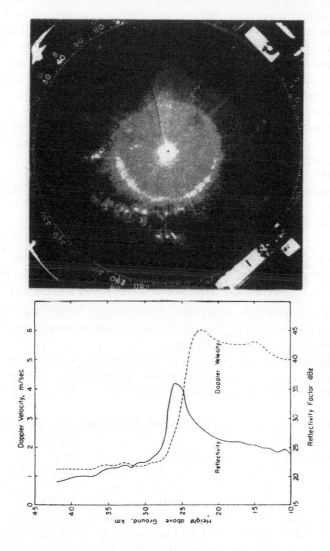

FIG. 12.3. Two views of the radar bright band: at the left a vertical profile of reflectivity and Doppler velocity as measured with a vertically pointing Doppler radar; at the right a PPI map at 8° elevation on which the melting layer appears as a bright ring at a range of about 20 km. (Doppler data from Cornell Aeronautical Laboratory; photograph from McGill Radar Weather Observatory.)

below or the snow above. The Doppler velocity in this profile is seen to increase from about 1.5 m/s in the snow to 6 m/s in the rain, with the maximum gradient in velocity located just below the level of maximum reflectivity. As snowflakes descend into air with a temperature warmer than 0°C, their radar reflectivity increases for several reasons, the most important of which is melting, because the dielectric constant of water exceeds that of ice by a factor of 4.4, or 6.5 dB. Also, in its initial stages, melting produces distorted wet snowflakes with somewhat higher reflectivities than those of spherical drops of the same mass. Continuing to melt while descending, the snowflakes become more compact and finally collapse into raindrops. Since the raindrops fall faster than the snowflakes their concentration in space is reduced. This dilution of the numbers accounts in part for the decrease of reflectivity in the lower part of the melting layer. If the melting flakes break apart further reduction in reflectivity occurs.

Wexler (1955) analyzed the processes in the melting layer and gave the following estimates of the changes in reflectivity arising from the different effects:

	Melting	Fall velocity	Shape	Condensation	Total
Snow to bright band	+6	−1	+1½	0	+6½ dB
Bright band to rain	+1	−6	−1½	+½	−6 dB

The fact that observations often reveal a stronger bright band than predicted suggests that aggregation in the upper part of the melting layer and disintegration below are occurring.

There are occasions in widespread precipitation when the bright band is weak, diffuse or entirely absent because of convective overturning: mixing disrupts the stratification necessary for the melting layer to be well defined.

Convection in widespread precipitation also manifests itself in so-called snow "generating cells". Figure 12.4 shows examples of measurements in snow with a radar using a fixed, vertically pointing beam. As the snow moved through the beam, the reflectivity was recorded as a function of time and altitude. In these coordinates, the data accurately portray the details of precipitation structure in the vertical. The observed time variations in data recorded this way arise from a combination of (1) the translation across the beam of spatial variations in the precipitation pattern and (2) evolution of the structure in time. The snow is widespread in all four examples, but cases (b) and (c) are more stratiform in appearance than the others. Patterns of type (a) and (d) are observed in advance of snow reaching the ground; most snowfall is associated with a trail pattern as in (b). These trails originate in compact cells called generating cells which are smaller than the trails and not often actually

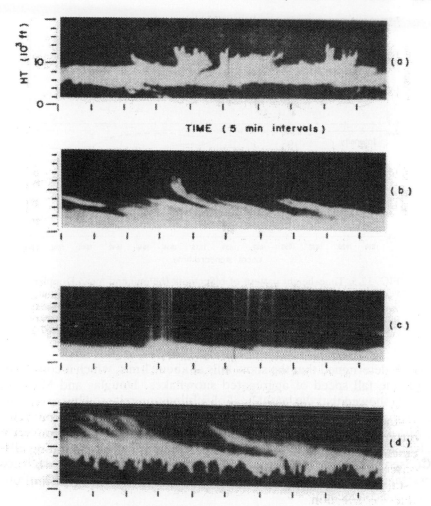

HT (10³ ft)

TIME (5 min intervals)

FIG. 12.4. Time–height records of (a) snow aloft, but not reaching the ground, (b) well-defined trail pattern, (c) relatively homogeneous echo, (d) stalactites (downward protrusions) at leading edge of storm. Total length of records is 50 min; vertical extent is 20,000 ft. (From Douglas *et al.*, 1957.)

observed. Pendulous extensions of the lower edge of the echo in (d) are referred to as stalactites, and occur when snow falls into dry air. Sublimation of the snow chills the air, causing local overturning that perturbs the lower echo boundary.

Snow generating cells are found to have no preferred altitude or temperature, but to be located usually just above frontal surfaces in air that is hydrostatically stable. From the slope of snow trails and knowledge of the speed of motion of the pattern through the beam, Marshall

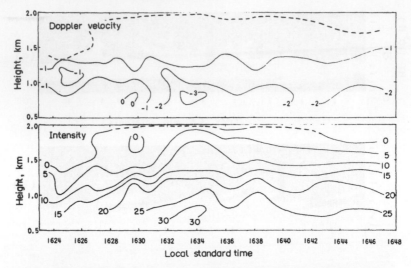

FIG. 12.5. Time–height pattern of reflectivity (below) and mean Doppler velocity (above) in Hawaiian orographic rain. The 0-dBz reflectivity contour, defining the top of the precipitation, is shown dashed in the Doppler velocity pattern. The vertical air velocities deduced from these observations were estimated to be everywhere less than 1 m/s. (Adapted from Rogers, 1967.)

(1953) determined that the snow falls at about 1 m/s, which is consistent with the fall speed of aggregated snowflakes. Douglas and Marshall (1954) showed that the latent heat of sublimation released by ice crystals growing in a moist stable environment is sufficient to initiate convective overturning, and that the vertical development of these convective elements is comparable to the observed dimensions of generating cells. Consequently the generating cells are probably regions in which ice crystal growth by aggregation and accretion is promoted by sublimation-induced convection.

Although widespread snow often presents the most stratiform appearance to radar of any precipitation, rain can also exhibit marked horizontal homogeneity. Figure 12.5 is an example of light rain produced by clouds consisting entirely of water droplets. The observations were obtained by a vertically pointing Doppler radar. The reflectivity and mean Doppler velocity were recorded as functions of time and altitude as the rain moved through the beam. In this example the pattern had a velocity of about 3 m/s. Therefore the record of 24 min duration corresponds to a distance through the rain pattern of about 4 km. Because the reflectivity contours are approximately horizontal, especially during the latter half of the record, it is reasonable to suppose that the rain pattern is slowly changing and that the figure approximately represents the rain pattern in space, along a line in the direction of its motion.

The horizontally stratified pattern, with reflectivity and downward Doppler velocity increasing progressively with distance downward from the top of the echo, indicates a steady precipitation process in which small raindrops near the echo top slowly descend through cloud and smaller raindrops, growing by coalescence. Such a process produces an increase in average drop size with distance fallen, with a corresponding increase in the reflectivity.

Showers

Two examples of PPI records of showers are shown in Fig. 12.6. Some of the showers in the first example are arranged in a line to the west of the radar, but in the second no particular organization is evident. Individual echoes in patterns of this sort have lifetimes of less than an hour, and the patterns themselves evolve rapidly. Convection instead of gravitational settling dominates the precipitation growth process.

Figure 12.7a shows the reflectivity structure of a rainshower as measured by a vertically pointing Doppler radar. The shower was moving through the beam with a velocity of about 4 m/s. Consequently the record, of 18-min duration, corresponds to a distance through the shower of about 4 km. The shower is thus of compact form, with approximately the same extent in the horizontal as in the vertical. Figure 12.7a does not depict the exact form of the shower in space, because development and internal changes were undoubtedly occurring during the observation time.

The Doppler velocity pattern of this shower (not shown) was used in connection with the reflectivity data to deduce the pattern of updraft velocity in Fig. 12.7b. At 1414 there was upward air motion through the entire vertical extent of the pattern, with maximum velocities exceeding 5 m/s. These upward motions include the 45 dBz maximum in the reflectivity pattern. At 1421 the signal again intensified overhead in connection with a region of vertical air motion, this time somewhat weaker than earlier. Heavy rain fell while the shower passed; total accumulation during the 18 min period was 3.8 mm.

The broad features of this example of a Hawaiian shower are relatively simple. It appears to consist of two active convective elements, each having a fairly continuous updraft and an associated region of high reflectivity. Most showers that have been observed in this manner appear to be multicellular. There are uncertainties in interpreting time-height records, however, for it is often not possible to determine whether the core of the shower or a fringe area is passing overhead.

Browning *et al.* (1968) reported on a shower that was observed simultaneously by two Doppler radars, one pointing vertically and the other at a low elevation angle. The center of the shower was known to

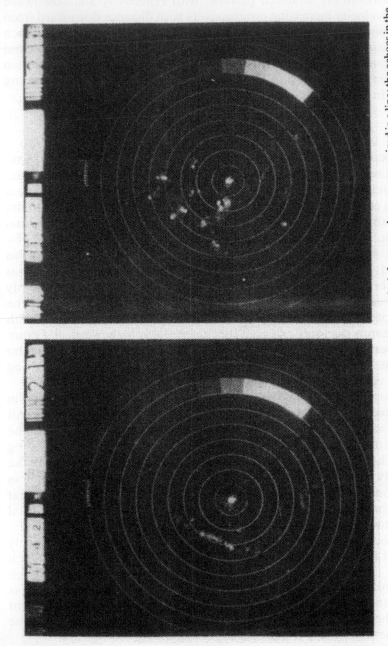

FIG. 12.6. Two examples of radar records of showers. In the picture on the left some echoes are organized in a line; the echoes in the right picture are located randomly. Range rings at 10-min intervals. Gray scale thresholds in 10-dB steps, with calibration pattern at 70 min to the east. (From Alberta Hail Studies Laboratory.)

pass directly over one of the radars during the observing period, and from the data it was possible to infer the pattern of air motion in the shower and the region of precipitation growth (Fig. 12.8). The updrafts are confined to the upper part of the shower (an observation in common with many other Doppler radar studies of showers) but are very weak, amounting to only about 1 m/s. Precipitation in the form of graupel is thought to originate and begin growing in this area. The graupel continues to grow as it descends through the weak updrafts and forms a precipitation streamer at lower levels. The solid trajectories in the figure are estimates of the paths of the largest graupel particles. Downdrafts predominate at low levels, again in common with other radar-measured airflow patterns. This rather characteristic pattern—updrafts at high levels and downdrafts below—suggests that the initial ascent of the air from low levels takes place outside any existing precipitation and is completed by the time the cloud particles in the ascending air have grown to radar detectable size. The lifetime of the convective element is thus about the same as the time required for the precipitation to develop. This fundamental characteristic of showers was pointed out by Houghton (1968), and will be mentioned later in the context of cloud modification.

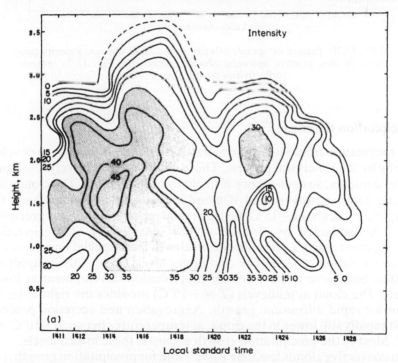

FIG. 12.7a. Time–height pattern of reflectivity in a warm-rain shower observed in Hawaii. Contours are labeled in dBz. Reflectivities greater than 30 dBz are shaded.

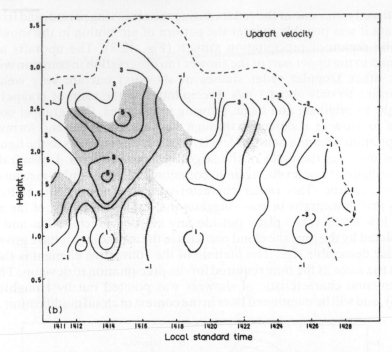

FIG. 12.7b. Pattern of updraft velocity in the same shower. Velocity contours in m/s, positive upwards. Shaded areas from Fig. 12.7a indicate reflectivities greater than 30 dBz.

Precipitation theories

Precipitation from midlatitude stratiform clouds is thought to develop chiefly by the ice-crystal process. These clouds have relatively low liquid water contents, so coalescence is likely to be ineffective. The clouds last a long time, however, and if cloud persists at altitudes where the temperature is about −15°C the ice-crystal process can lead to precipitation. As explained by Braham (1968) in a survey of precipitation development, each level in stratiform clouds has a special importance in the precipitation process. The cold upper levels ($T \approx -20°C$) supply ice crystals that serve as embryos for precipitation development at lower levels. The cloud at midlevels ($T \approx -15°C$) provides the right environment for rapid diffusional growth. Aggregation and accretion proceed most rapidly still lower in the cloud, at temperatures between −10°C and 0°C. Most of the precipitation growth occurs in these lowest levels.

In convective clouds less time is available for precipitation growth, but because liquid water contents are typically higher than in stratiform clouds coalescence stands a better chance of producing rain. From the

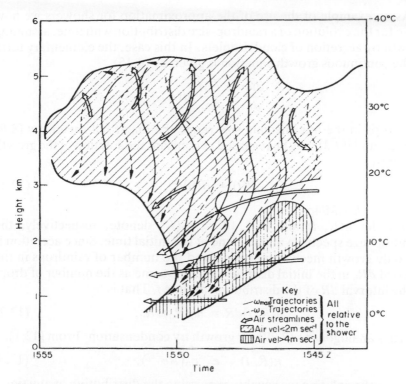

FIG. 12.8. Time–height pattern of a shower showing streamlines of air motion, trajectories of the largest particles (ω_{max}), and trajectories of the particles contributing most to the Doppler velocity (ω_p). The streamlines and trajectories are drawn relative to the shower. Temperature profile indicated on the right. (From Browning *et al.*, 1968.)

observation that the lifetime of a convective element (about 20 min) is also the time needed for precipitation to grow, Houghton (1968) concluded that the precipitation forming process must begin early in the developing cloud and therefore at a low level. Although the precipitation may be initiated by either coalescence or the ice-crystal process, depending primarily on the temperature and cloud water content, most precipitation growth is by accretion.

Thus the mechanisms of precipitation formation are quite different in stratiform and convective clouds. As a useful approximation continuous rain can often be viewed as a steady-state process, in which cloud quantities may vary with height but are constant with time at any given height. On the other hand, showers may be approximated as systems in which the cloud properties vary with time but are constant with height at any given time. These limiting approximations were first suggested by Rigby, Marshall, and Hitschfeld (1954).

As an example of the use of the approximation for showers we now solve for the evolution of a raindrop-size distribution with time, assuming growth by accretion of cloud droplets. In this case, the elementary form of the continuous growth equation is given by (8.15),

$$\frac{dR}{dt} = \frac{\bar{E}M}{4\varrho_L} u(R).$$

For drops in the intermediate size range the linear fall speed law (8.8) applies, and if $\bar{E}M$ may be regarded as constant the solution of the growth equation is

$$R(t) = R(0)e^{at}, \tag{12.1}$$

where $a = k_3\bar{E}M/4\varrho_L$.

Now let $n(R, t)$ and $n(R, 0) = n_0(R)$ denote, respectively, the raindrop-size spectrum at time t and at the initial time. Since accretion is the only growth mechanism considered the number of raindrops in the interval dR_0 in the initial distribution is the same as the number of drops in the interval dR of the distribution at time t. That is,

$$n(R, t)dR = n_0(R_0)dR_0, \tag{12.2}$$

which is analogous to (7.29) for growth by condensation. From (12.1),

$$n(R, t) = e^{-at}n_0(Re^{-at}), \tag{12.3}$$

which is the solution we sought, expressing the distribution at any time t in terms of the initial distribution. This approximation neglects coalescence among raindrops, and is probably most appropriate for the early stages of raindrop growth in convective clouds, when light rain is falling through relatively dense cloud.

In the continuous-rain approximation the number flux of raindrops is constant with height; otherwise the drop-size distribution would vary with time. Therefore the initial distribution and the distribution after distance of fall h are related by

$$n(R, h)u(R)dR = n_0(R_0)u(R_0)dR_0. \tag{12.4}$$

Assuming negligible updraft speed and again employing (8.8) for the fall speed, we find

$$n(R, h) = \left(1 - \frac{bh}{R}\right)n_0(R - bh), \tag{12.5}$$

where $b = \bar{E}M/4\varrho_L$.

Though only coarse approximations, (12.3) and (12.5) give an indication of the difference between the two idealized precipitation processes. In effect, these results represent extensions of the continuous-growth accretion equations to drop populations.

As an example of the usefulness of these approximations, let us solve for the rate at which precipitation is produced in a developing shower. The rainwater content at time t is given by

$$L(t) = \frac{4}{3}\pi\varrho_L \int_0^\infty n(R)R^3 dR.$$

From (12.2), for the shower process, we can express L in terms of the initial drop-size distribution:

$$L(t) = \frac{4}{3}\pi\varrho_L \int_0^\infty n_0(R_0)R^3 dR_0.$$

If now we use the linear fall speed assumption, R and R_0 are related by (12.1) and we have

$$L(t) = \frac{4}{3}\pi\varrho_L \int_0^\infty n_0(R_0)R_0^3 e^{3at} dR_0 = e^{3at}L_0,$$

where

$$L(t) = \frac{4}{3}\pi\varrho_L \int_0^\infty n_0(R_0)R_0^3 dR_0$$

is the initial rainwater content. Thus, if $\overline{E}M = 1 \text{ g/m}^3$, for example, then $a = 2.0 \times 10^{-3} \text{ s}^{-1}$ and the liquid water content doubles in a time equal to $(\ln 2)/3a = 116 \text{ s} = 1.93 \text{ min}$.

Mesoscale structure of rain

Many of the roots of modern meteorology can be traced to the work of a group of Scandinavian researchers under the leadership of Vilhelm Bjerknes in the years around 1920. This group, called the Bergen school, invented synoptic meteorology as we know it today. In developing the theory of air masses, cyclones, and fronts, they did not fail to see that the spatial distribution of precipitation was associated with the synoptic pattern, and that the precipitation itself had a characteristic pattern on a smaller scale. One of the group members, Tor Bergeron, who was later to become a leader in cloud microphysics, studied squall lines and other structural details of precipitation in relation to the fronts and cyclones. Friedman (1982) may be consulted for a review of the scientific accomplishments of the Bergen school.

Following World War II, mainly because of the advent of radar, it has become possible to observe the structure of precipitation in greater detail than ever before. Satellites have helped to place these observations into a larger perspective, and improved airborne instruments for cloud physics research have filled in the microphysical details. The word mesoscale was coined for meteorology to describe the new detail observable by radar. This subject merits attention because much of the important weather that affects our activities—heavy rain, high winds, snowstorms—is mesoscale weather, not resolvable in the synoptic network on which forecasts are based. The investigations of mesoscale precipitation structure have consisted of attempts to classify recurring features and of efforts to understand the links between dynamics and microphysics that explain the structure.

Heavy rain is often observed to be arranged in lines or bands. Figure 12.9 is a radar example of showers and non-severe thunderstorms spread over a line about 400 km long. The line corresponds closely to the position on the surface of a stationary front. Convection is triggered by the front and the echoes mark its presence. Depending on the airflow, stability, moisture supply, and orographic influences, rain may occur in advance of a front, behind it, or not at all, and it may or may not show a

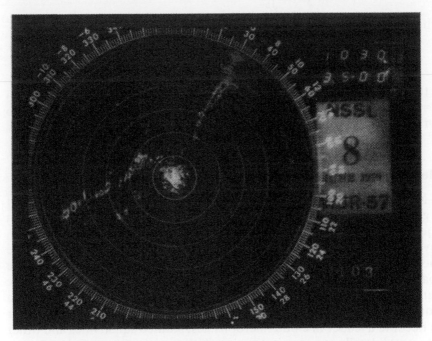

FIG. 12.9. Showers and thunderstorms located along a stationary front oriented NE–SW. Radar located at National Severe Storms Laboratory, Norman, Oklahoma. Range rings at 40-km intervals. These are analog data displayed on a scale of three shades. The first shade, defining the outer boundary of the echoes, corresponds to 20–25 dBz. The second and brightest shade extends from 25 to 35 dBz. The next interval, 35–45 dBz, is reversed to appear black. Most of the echoes along this line appear to have "holes", which indicates that their core reflectivities are in the 35–45 dBz range. (Courtesy of NSSL.)

banded structure. Convective rain can also develop in suitably unstable air without frontal lifting. Not uncommonly, as it develops, the rain then becomes organized in bands or lines unrelated to the synoptic pattern.

Mesoscale organization may exist in the form of patterns other than lines or bands. Figure 12.10 is a sequence of radar maps at one-hour intervals showing the development of a large but compact area of heavy rain. Initially, there is a mass of mixed convective and stratiform rain east of the radar. The more intense echoes are aligned along the southeastern edge of this mass, but heavy rain extends over an area at least 200 km long and 150 km wide. As this area moves off to the east, a line of new convective storms forms to the southwest. The line moves east, while developing on its northern side and filling in, until at the end of the sequence it has about the same shape and location as the heavy rain area 6 h earlier. The reasons for the shape of this pattern and the time scale of its recurrence are not known in any detail. Although heavy rain was expected from the synoptic conditions, its timing and location, which are mesoscale considerations, could not be predicted accurately more than a short time in advance.

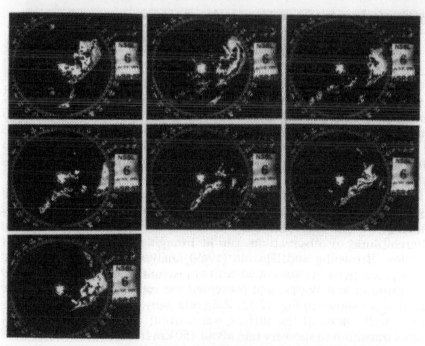

FIG. 12.10. Sequence of radar maps (left-to-right, from the top down) at intervals of approximately 1 h, from 1948 to 0140 local time, June 6–7, 1979. The same display convention as in Fig. 12.9 is used, a staggered gray scale. In these echoes, two higher levels appear: 45–55 dBz, which is displayed dim, and >55 dBz, shown bright. (Courtesy of NSSL.)

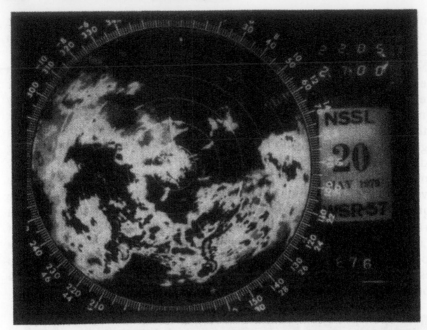

FIG. 12.11. Plan view of chaotic precipitation pattern. Close inspection reveals a few small areas with reflectivities greater than 55 dBz. (Courtesy of NSSL.)

It is wrong to think that heavy rain always has a pattern with recognizable structure. Figure 12.11 is an example of mixed convective and widespread rain following a day of severe weather. A short (50 km) band of intense rain is located 150 km south-southeast of the radar, but otherwise the pattern is chaotic, bearing no simple relation to the synoptic map.

In spite of the complexity of the problem there has been progress in the classification of mesoscale precipitation, and to some extent in understanding it. The approach used is intensive case studies, in which different kinds of observations can be brought to bear on a particular situation. Browning and Harrold (1969) analyzed the air motion and precipitation patterns associated with an occluding cyclone that passed over England and Wales, and presented the schematic distribution of rainfall types shown in Fig. 12.12. Rain relatively uniform in pattern was located well ahead of the surface warm front, and there was a fairly distinct transition to showery rain about 150 km from the front. Bands of showers in the warm sector were orographically influenced and aligned parallel to the winds at about 700 mb. These bands extended ahead of the warm front, merging with smaller bands aligned with the front. The low-level air ahead of the warm front was convectively unstable. The

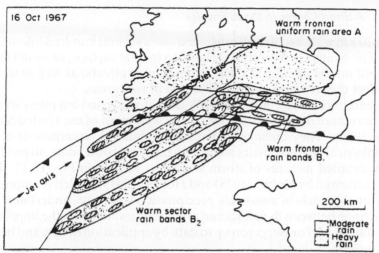

FIG. 12.12. Schematic diagram showing the distribution and structure of rainfall in a cyclone wave. The axis of the high level jet stream is shown passing directly over the Irish Sea. (From Browning and Harrold, 1969.)

changing character of the rain was a result of ascent at the surface warm front, which eventually triggered the instability.

Using radar observations and raingauge records, Austin and Houze (1972) studied the precipitation patterns of nine New England storms covering a wide range of seasonal and synoptic situations. Although at first glance quite dissimilar, all patterns were found to be composed of subsynoptic-scale precipitation areas with rather clearly definable characteristics. These areas could be grouped into four categories: synoptic areas which are larger than 10^4 km^2 and have a lifetime of one day or longer; large mesoscale areas which range from 10^3 to 10^4 km^2 and last several hours; small mesoscale areas which cover 100–400 km^2 with a lifetime of about an hour; and smaller elements which are about 10 km^2 in size and last usually no longer than half an hour. In all cases studied it was found that every precipitation area of any of these scales contained one or several of each of the smaller sized areas. An investigation of the precipitation intensities within the various areas showed that the rainfall rates in large mesoscale areas were 2–4 times greater than those on the synoptic scale; that the rain rates in the small mesoscale areas were about double those in large mesoscale areas; and that the rain rates in the smallest elements were 2–10 times those in the small mesoscale areas. Although the smallest elements have the highest rain rates, the main contribution to the total rainfall on the synoptic scale comes from the small and large mesoscale areas.

Summarizing studies of this kind, Harrold and Austin (1974) noted that regions of heavy rain can be found in widespread rain as well as showery situations and tend to occur in compact groups rather than to be randomly scattered. The groups are often in the form of bands, typically about 20 km wide and in well organized cases several hundred kilometers long. The bands are sometimes related to frontal surfaces or squall lines, but need not be parallel to them. Topographic effects, as well as fronts, can affect the structure and development of rain areas.

Researchers at the University of Washington carried out many studies of the structure of cyclonic storms off the west coast of the United States in the 1970s. They included extensive airplane observations of cloud microphysical characteristics along with radar and other data, to produce highly detailed pictures of storm structure, such as in Fig. 12.13. This work, reviewed by Hobbs (1978) and Houze (1981), has led to improved conceptual models of mesoscale precipitation structure, underlining the interactions between dynamics and microphysics. It brings the important work started by Tor Bergeron up to date by application of new and better observations.

Seltzer *et al.* (1985) sought an explanation for the formation of rainbands by mechanisms other than frontal lifting. They selected for analysis 15 cases of rainbands in New England in which the surface low center was at least 500 km away from the rainband and there were no surface fronts nearby. Figure 12.14 is a typical example, with a rainband oriented in the north–south direction, parallel to the shear of the geostrophic wind. They assessed the possible importance of symmetric instability using the criterion following from (3.23), namely

$$\frac{\delta z}{\delta y} > \frac{f - \dfrac{\partial u_g}{\partial y}}{\dfrac{\partial u_g}{\partial z}}, \tag{12.6}$$

where $\delta z/\delta y$ is the slope of the isentropic surface and the right-hand side represents the slope of the absolute vorticity vector.

By writing $\delta z/\delta y$ as $-(\partial\theta/\partial y)/(\partial\theta/\partial z)$, employing (3.19), and introducing the Brunt–Väisälä frequency from (3.14), we can change (12.6) to

$$\frac{f \cdot (\partial u_g/\partial z)}{N^2} > \frac{\eta}{(\partial u_g/\partial z)},$$

where $\eta = (f - \partial u_g/\partial y)$ is the vertical component of the absolute vorticity. A further simplification can be made by introducing the Richardson number, defined by

$$Ri = \frac{N^2}{\left(\dfrac{\partial u_g}{\partial z}\right)^2},$$

so that the condition for symmetric instability becomes

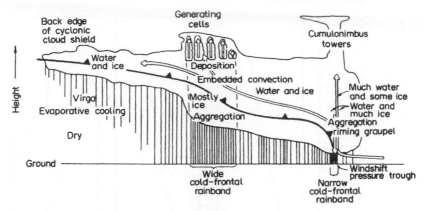

FIG. 12.13. Organization of precipitation around a cold front. (From Hobbs, 1978.)

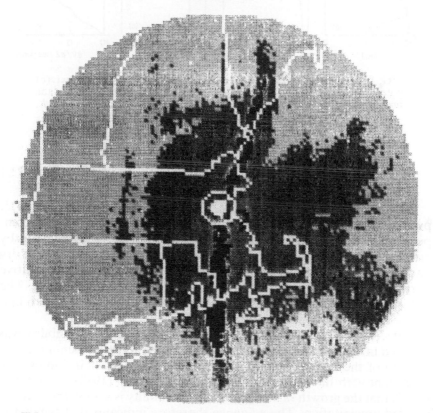

FIG. 12.14. PPI radar map showing the average reflectivity factor in a layer from 2.5 to 4.0 km altitude at 0000 GMT, December 6, 1981. The two shades correspond to reflectivities greater than 15 dBz and 25 dBz. (From Seltzer *et al.*, 1985.)

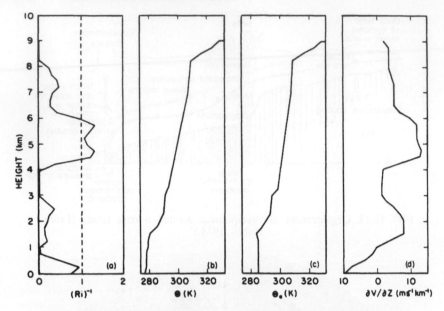

FIG. 12.15. Vertical profiles of (a) reciprocal of the Richardson number, (b) potential temperature, (c) equivalent potential temperature, and (d) the magnitude of the wind shear parallel to the rainband. (From Seltzer *et al.*, 1985.)

$$\frac{\eta}{f} < (Ri)^{-1}. \tag{12.7}$$

Figure 12.15 from Seltzer *et al.* shows vertical profiles of $(Ri)^{-1}$, potential temperature, equivalent potential temperature, and the component of the wind shear parallel to the rainband, all from a nearby atmospheric sounding. The atmosphere is seen to be stable for both dry and moist adiabatic ascent. The airflow was such that the relative vorticity was much smaller than f, so that $\eta/f \approx 1$ at all altitudes. The criterion (12.7) therefore indicates that the layer between 4 and 6 km, where $Ri < 1$, is symmetrically unstable.

Seltzer *et al.* showed that the condition for symmetric instability was satisfied in all 15 cases analyzed, and that the predicted orientation and spacing of the rainbands agreed with observations. Their findings are consistent with the analysis of Bennetts and Sharp (1982), who determined that the growth rate of symmetric instability is a good predictor of banded precipitation. The evidence therefore indicates that symmetric instability may be responsible for the formation of midlatitude rainbands that are not related to frontal lifting or topographic effects.

Precipitation efficiency

Clouds provide the intermediate step in converting atmospheric water vapor to precipitation. Not all rainclouds are equally effective in accomplishing this conversion. Small cumulus clouds, for example, often grow rapidly but begin to dissipate just as precipitation develops. Consequently much of the cloud water is not converted to precipitation but remains aloft, eventually to evaporate. For a different reason many stratiform clouds are also ineffective in producing precipitation. Although they may last for hours, they have neither the high liquid water contents which favor coalescence nor the cold temperatures needed to initiate the ice crystal process. Therefore little precipitation occurs even though the cloud may be supercooled aloft, and hence microphysically unstable with respect to the ice crystal process. The concept of precipitation efficiency has been used to describe, from several points of view, how effectively a cloud converts either vapor or condensed material to precipitation.

Braham (1952) determined the water budget of small thunderstorms by analyzing extensive data on clouds in Florida and Ohio. For the average inflow of water vapor into these storms he found 8.9×10^8 kg. Of this amount, 5.3×10^8 kg is condensed; the remainder leaves the storm without condensing. Of the water that condenses, only about 10^8 kg reaches the ground as rain. The rest evaporates in the downdraft or at the sides of the cloud. The precipitation efficiency may be defined as the ratio of the mass of rain reaching the ground to the mass of vapor entering the cloud. With this definition the efficiency is only 11%. If the efficiency is defined as the fraction of condensed water that ultimately reaches the ground, it becomes 19%, a figure often quoted for thunderstorms.

Wexler (1960) defined the precipitation efficiency somewhat differently, as the ratio of the amount of precipitation that falls to the water made available by condensation in pseudoadiabatic ascent. He found that the most efficient clouds in this sense are cumulus embedded in widespread stratus.

Looking at precipitation efficiency from the microphysical point of view, Hardy (1963) made calculations with different forms of the raindrop size distribution $N(D)$ to determine which were the more effective in depleting (sweeping out) the cloud droplets. He concluded that steep distributions (large values of Λ in the exponential formula 10.1) were the most effective in this sense. From radar observations in Hawaiian orographic rain, which is characterized by numerous small drops, Rogers (1967) found support for Hardy's idea, noting that the exceptionally rapid decrease of reflectivity with height implies efficient sweepout.

The sweepout efficiency was examined by Houghton (1968) in an appraisal of schemes for artificial precipitation modification. He defined the efficiency S by

$$S = \frac{\pi}{4} \int N(D)D^2V(D)dD, \tag{12.8}$$

where the integration extends over all drops, assumed spherical. Physically, S is the fraction of a unit horizontal area in the cloud that is geometrically swept out by precipitation particles in unit time. From data on drop-size distributions, Houghton determined that to a good approximation $S \propto R$. He found that the "scavenging" is incomplete for rain-showers, which is one cause of their low precipitation efficiency (Braham's 19%). With regard to the feasibility of modifying precipitation, Houghton concluded that owing to the inefficiency of showers opportunities exist for stimulating rain by cloud seeding, but these opportunities occur only under certain specific conditions and at particular times.

Acidic precipitation

Mounting concern over the apparent widespread environmental effects of acidic precipitation has focused attention on the long neglected subject of atmospheric chemistry. In broad outline, the formation of acidic precipitation is understood to consist of the following steps:

1. Emission into the atmosphere of acid precursors, mainly oxides of sulfur and nitrogen, species that occur naturally but also as by-products of fuel combustion and other industrial activities.
2. Transport of the pollutants by atmospheric motions.
3. Transformation of the pollutants by chemical and physical processes, such as oxidation, hydrolysis, and coagulation.
4. Deposition of the pollutants on the earth's surface in precipitation as a dilute solution of sulfuric and nitric acid.

Clouds are an important link in the so-called "acid rain" phenomenon not only because they are the source of the precipitation, but also because they are the medium in which major reactions occur that lead to the formation of acidic substances.

A common measure of the acidity of a solution is the hydrogen ion concentration, usually expressed in terms of $pH = \log(1/[H^+])$ where $[H^+]$ is the hydrogen concentration in moles per liter. Pure water has a pH of 7. If acid is added, the hydrogen content will increase and the pH will therefore decrease. Natural cloud water and precipitation reaches an equilibrium with atmospheric carbon dioxide to form weak carbonic acid with $pH = 5.6$ at sea level. By convention, precipitation with $pH < 5.6$ is designated acidic. Naturally occurring sulfate (SO_2), reacting with trace gases, tends to reduce the pH even further, so 5.0 might be a more appropriate "background" against which to look for pollution effects. The natural background is actually highly variable, depending on the presence of local sources of chemically active materials, such as basic soil particles, which can affect the acidity of precipitation. Two large areas of the world have precipitation with average pH values less than 5 and core values near 4. These are located over western Europe and eastern North America. Both are in or just downwind of major industrial areas that together account for about one-third of the total of man-made emissions. Non-acidic precipitation, with pH values ranging as high as 7.5, is reported for large continental interiors and remote Arctic locations. However, for reasons unknown, some sites far from obvious pollution sources have pH values as low as about 4.5.

Two chemical pathways are recognized whereby sulfur dioxide and oxides of nitrogen can be transformed to acidic precipitation. Both require oxidation as a step in the formation of highly soluble acids. One path consists of gas-phase oxidation followed by uptake in cloud or rainwater. The other consists first of dissolution of the gas followed by aqueous-phase oxidation. These reactions are strongly influenced by the presence of oxidizing trace gases such as ozone (O_3) and hydrogen peroxide (H_2O_2). Atmospheric ammonia (NH_3) in trace amounts tends to neutralize the acidity somewhat.

Gas-phase reactions can lead to the formation of gaseous sulfuric acid (H_2SO_4), which readily converts to an aerosol form by combining with existing aerosols or by nucleating with water vapor. Nitric oxide (NO) rapidly oxidizes to nitrogen dioxide (NO_2) in the presence of trace amounts of O_3. Gas-phase oxidation of NO_2 then leads to nitric acid gas, HNO_3, which is quickly taken up by the liquid water in clouds to form aqueous nitric acid.

The aqueous-phase path is potentially important because a cloud of liquid droplets is a favorable medium for aqueous reactions with gases. The significant feature of clouds is that the water is highly dispersed, and hence presents a high surface-to-volume ratio for the uptake of acidic species. This can lead to high concentrations when gaseous species are dissolved in the water. Laboratory studies indicate that the most significant aqueous-phase reactions are the oxidation of SO_2 by H_2O_2 and O_3. The rate of reaction with O_3 is self-limiting and becomes unimportant when the pH is below about 4.5. The reaction with H_2O_2 is therefore thought to be an explanation of the production of cloud and rainwater having pH < 4.5.

Some of the outstanding questions in atmospheric chemistry concern the rates of production and destruction of the trace gases. For example, hydrogen peroxide appears to be necessary for the aqueous-phase oxidation leading to H_2SO_4, but it is consumed by this process. Evidently the H_2O_2 is produced by gas-phase reactions that require the hydroxyl radical (OH), itself a photochemically produced trace species. Cloud models are being developed that include the coupled chemistry of trace gases with the processes of droplet growth and precipitation development. An important area of research is the coordination of modeling efforts with experimental investigations of emissions, transports, and cloud chemistry. Recommended reviews of this complex subject are those of Summers (1982), Morgan (1982), and Schwartz (1987).

Problems

12.1. The sweepout efficiency of a single precipitation particle may be defined as the volume of space that it geometrically sweeps out per unit time. Compare the sweepout efficiency of a graupel particle with that of a small raindrop having the same mass. For graupel, assume the relations between mass, radius, and terminal fall velocity as given in problem 10.2. For the raindrop, assume the linear fall speed relation, $u(r) = k_3 r$, where $k_3 = 8 \times 10^3 \, s^{-1}$.

12.2. In a rapidly developing cumulonimbus cloud, the radar echo is observed to appear simultaneously over a deep interval of altitude. For the particular radar used, the initial echo is caused by the formation of drizzle-size drops. As the cloud continues to develop, these drops grow rapidly by sweeping out cloud droplets, and the radar echo strengthens at all levels.

Calculate the time required for the signal to increase by 10 dB. Make the following assumptions:

(1) Cloud liquid water content M is constant at 5 g/m^3.
(2) All the precipitation growth is by the collection of cloud droplets; there is no growth by diffusion or by coalescence among the raindrops.
(3) The effective average collection efficiency is 0.5.
(4) Raindrop fall speed may be approximated by the linear formula of problem 12.1.
(5) Raindrops account for the reflectivity; the contribution of cloud droplets is negligible.
(6) Breakup effects are negligible.

12.3. The vertical gradient of reflectivity in stratiform rain gives an indication of the extent to which raindrops are growing by sweeping out cloud droplets. A strong decrease of Z with height (or increase with distance fallen) indicates rapid accretional growth.

In the lowest 1 km of nimbostratus cloud, the raindrops are growing by accreting cloud droplets. The effective cloud liquid water content (the product of E times M) equals 2 g/m^3. Using the elementary form of the continuous growth equation, show that the reflectivity increases by approximately 19 dB in the lowest kilometer of the cloud. Solve also for the fractional increase of rain liquid water content, L, in the lowest kilometer of the cloud. Make the following assumptions:

(1) steady-state process;
(2) zero updraft velocity;
(3) drop growth by accretion only;
(4) linear fall speed dependence appropriate for small raindrops;
(5) raindrop size distribution at 1 km above cloud base of Marshall–Palmer form corresponding to $R = 0.1$ mm/h.

12.4. As an index to the efficiency of precipitation growth by accretion, H. G. Houghton (1968) considered the total fraction of a unit horizontal area in a cloud that is geometrically swept by the precipitation particles per unit time. Derive an expression for this efficiency as a function of rainfall rate for the following precipitation model:

   (i) Drop-size distribution of the general exponential form $N(D) = N_0 \exp(-bD)$ with N_0 a constant equal to 0.08 cm^{-4} and b a parameter depending on the rate of rainfall.
   (ii) Terminal velocity related to size by $u(D) = kD$, with $k = 4 \times 10^3$ s^{-1}.
   (iii) Zero vertical air velocity.

Evaluate this expression for a rainfall rate of 10 mm/h.

12.5. Develop an expression for the rate at which cloud water content M is depleted by rain-collection. As in problem 12.4, assume a general exponential form, of $N(D)$ and the linear dependence of fall speed on drop size. Show that in a stagnant cloud, that is, one in which there is no vertical air motion and no change in cloud water content except by rain-collection, rain falling at a steady state of 10 mm/h for 5 min will reduce M to approximately 45% of its original value.

12.6. In the theory of the radar bright band, one of the effects considered is the increase of reflectivity within the melting layer caused by condensation on the surfaces of the melting snowflakes. Show that approximately 60 g of water is condensed for each kilogram of snow that melts.

12.7. From a large sample of radar echoes at an altitude of 2 km in the vicinity of Montreal, a systematic relationship was found between the average echo size and the core reflectivity. The mean area of echoes defined by reflectivity threshold ζ (in dBz) that contain interior reflectivities as strong as ζ_i (dBz) is given approximately by

$$A_{\zeta_i}(\zeta) = A_0 \exp(\lambda\zeta_i - \gamma\zeta),$$

where $\gamma = 0.15$ (dBz)$^{-1}$, $\lambda = 0.14$ (dBz)$^{-1}$, and $A_0 = 150$ km^2. This approximation is valid for ζ between 15 and 50 dBz and for ζ_i between 30 and 50 dBz; it only has meaning if $\zeta < \zeta_i$.

Assume that the reflectivity factor is related to rainfall rate by

$$Z = 200R^{1.6}.$$

Use the defining equation for ζ,

$$\zeta = 10 \log_{10} Z \ (\text{mm}^6/\text{m}^3),$$

and solve for the area-average rainfall rate $\bar{R}$ in the echoes, in terms of R_{min} and R_{max}, where R_{min} is the rain rate corresponding to ζ and R_{max} is the rain rate corresponding to ζ_i. For $\zeta = 20$ dBz and ζ_i ranging from 30 to 50 dBz, show that $\bar{R}$ is within 40% of the value 2 mm/h.

13

Severe Storms and Hail

IN air that is sufficiently moist and unstable, convective clouds can grow to great heights, develop vigorous updrafts, and produce heavy rain, lightning, and hail. These large severe storms may occur individually or, more typically, in groups associated with synoptic-scale fronts or meso-scale convergence areas. In many parts of the world they are the cause of serious flooding, wind and hail damage, and loss of life.

Although our understanding of such storms is incomplete, extensive studies over the past three decades by means of radar, radiosonde networks, and instrumented airplanes have made it possible to describe their structure and development and to recognize the meteorological conditions under which they are likely to occur. This chapter describes the structure of thunderstorms and outlines the theory of hail growth.

Life cycle of the thunderstorm cell

From detailed observations of thunderstorms in Florida and Ohio, Byers and Braham (1949) found that the storms are made up of one or more units of convective circulation, consisting of an updraft area and a region of compensating downward motion. These convective cells are much the same in structure and behavior in most storms and may therefore be considered as a class of convective phenomena unique to thunderstorms. Often a cloud is made up of a number of cells in various stages of development, and it is difficult to identify any individual cell. However, it is convenient to consider the thunderstorm cell as the elementary unit of storm structure.

The life cycle of a cell is divided into three stages (see Fig. 13.1) depending on the predominant direction and magnitude of the vertical air motion:

1. Cumulus stage—characterized by an updraft throughout most of the cell.
2. Mature stage—characterized by the presence of downdrafts and updrafts.
3. Dissipating stage—characterized by weak downdrafts throughout most of the cell.

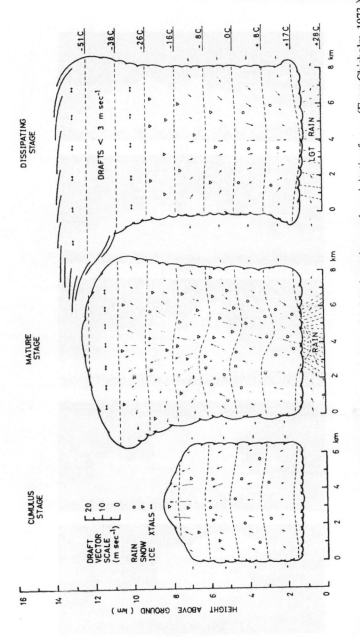

FIG. 13.1 The Byers–Braham model of a thunderstorm cell, indicating air motions and precipitation forms. (From Chisholm, 1973.)

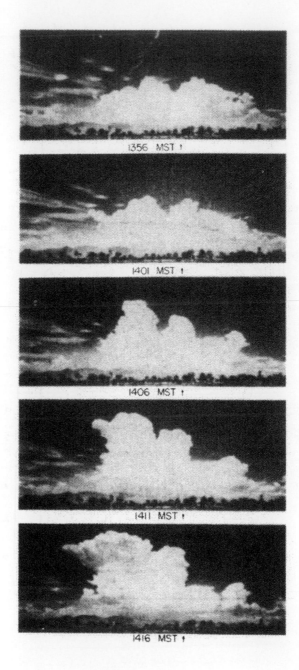

1356 MST ↑

1401 MST ↑

1406 MST ↑

1411 MST ↑

1416 MST ↑

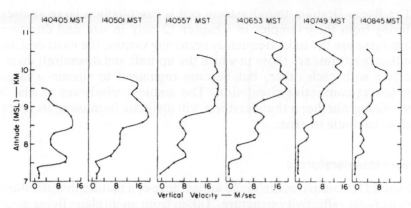

FIG. 13.2. A developing cumulonimbus cloud and radar-measured updraft velocities. The pictures (opposite) indicate two cells developing simultaneously. The measurements were obtained by a vertically-pointing Doppler radar located under the cloud at the position of the arrow. (From Battan and Theiss, 1966.)

As the updraft causes the cloud to grow in the cumulus stage, air flows in through the sides or the top ("entraining") and mixes with the updraft. With continued upward motion a large amount of water condenses and eventually falls as precipitation. This falling water initiates the downdraft because of viscous drag of the water on the air and evaporative cooling of the air. This is the start of the mature stage of development. The air of the downdraft reaches the ground as a cold core in the rain area and spreads over the surface, changing the surface wind pattern.

The downdraft interferes with the updraft at low levels in the cloud, and eventually cuts off the updraft from its source region. The cell then enters its dissipating stage. With the decay of the updraft and consequent elimination of the source of rainfall the downdraft weakens and finally dies out completely, leaving a residue of cloudy air.

The cumulus stage typically has a duration of 10–15 min. The mature stage lasts 15–30 min; though difficult to specify definitely, the dissipating stage lasts about 30 min.

Since the time of Byers and Braham's study, long-term field programs to investigate severe storms have been mounted and more sophisticated techniques have been developed for the observations. It has been found nevertheless that the relatively simple model of Fig. 13.1 adequately describes the general features of small, "single-cell" cumulonimbus clouds and of the convective elements in larger "multicell" clouds. Figure 13.2, for example, shows the vertical air velocity measured by a Doppler radar during the early mature stage of a growing cumulonimbus. Upward velocities are consistent with those in Fig. 13.1. The downdraft is not yet apparent, possibly because it was displaced out of the observing plane of the radar.

The Byers–Braham thunderstorm cell is essentially a large shower, differing from the examples in Chapter 12 only in size and duration. While these are the most frequently occurring storms, the most destructive thunderstorms are those in which the updraft and downdraft do not interfere with each other, but become organized to sustain a large, long-lasting convective circulation. The ambient winds are crucial in determining whether a thunderstorm will dissipate because of its downdraft or continue to exist.

Severe thunderstorms

Figure 13.3 compares the visual appearance of a mature thunderstorm with its radar reflectivity structure. Taken from an airplane flying south of the storm, the picture indicates the low and midlevel clouds associated with the storm, its intense central region or core, and the high level anvil cloud reaching off to the northeast. Air moving generally from the south at low levels is drawn into the updraft, supplying the moisture and buoyancy needed to sustain the storm circulation. A closer inspection of the photograph shows a newly developing mass of cloud turrets at low levels on the near side of the storm, which we can associate with the small radar echo to the south of the main body of the storm. Large thunderstorms often consist of several convective elements in different stages of development. The storm system then has a longer lifetime than the individual elements.

The core of the storm has a reflectivity factor exceeding 50 dBz, arising from a combination of heavy rain and hail. The total areal extent of the outermost (20 dBz) contour is about 240 km^2. The echo extends vertically throughout the troposphere to a height of 30 thousand feet, or nearly 10 km, and is limited by the stable stratosphere into which it has penetrated. The more active part of the storm is its southwestern side. This is where the new clouds and the small echo appear, where the horizontal reflectivity gradients are the strongest, and where an indentation appears in the main echo body. The indentation is interpreted as the region of main low-level inflow to the storm updraft. Such indentations are not always observed, but may be seen also in Fig. 13.4, which shows a pair of Alberta thunderstorms on another day at three elevation angles, corresponding to altitudes of approximately 2.8, 5.6, and 8.4 km above the ground.

The examples of Figs. 13.3 and 13.4 are severe hailstorms that are larger and longer lasting than the basic thunderstorm cell of Fig. 13.1. Because of the change of the environmental wind with height, the updraft and downdraft are horizontally displaced from one another and can interact mutually to sustain a strong, long-lived circulation. First analyzed by Browning and Ludlam (1962), this kind of storm circulation has become known as a "supercell".

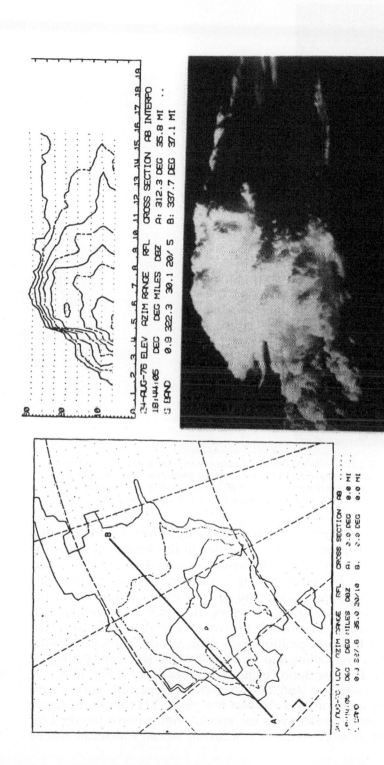

FIG. 13.3. An Alberta thunderstorm with two simultaneous radar patterns, one (at left) a low-elevation plan view, the other (above) a vertical cross section in an approximate northeast–southwest direction (line AB) through the storm core. The outermost contours correspond to 20 dBz. The contour spacing is 10 dBz on the PPI and 5 dBz on the vertical cross section. (Photo and computer-generated radar data courtesy of Dr R. G. Humphries, Alberta Research Council.)

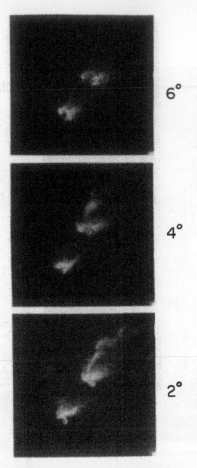

6°

4°

2°

FIG. 13.4. A pair of Alberta thunderstorms observed by radar at elevation angles of 2, 4, and 6 degrees. The range rings are at 10-mile intervals. These storms are located at a distance of about 80 km in the northwest quadrant of the radar map. They have the characteristics of supercell storms.

Figure 13.5 is a schematic diagram showing horizontal sections of a supercell echo at three altitudes. The key to this kind of storm is the ambient wind and its variation with height, indicated here by the vectors labeled L, M, and H, corresponding to the wind at low, medium, and high levels in the troposphere. The updraft enters at low levels and ascends in the region called by Browning the "vault". The updraft is so strong that precipitation is not able to grow to radar detectable size in the vault region. When precipitation does form at higher levels the wind shear prevents it from falling into the updraft at low levels and cutting off the circulation. This circulation is shown in more detail in Fig. 13.6.

Because of their size and destructiveness, supercell storms have received much attention over the past few years. Yet they occur rather infrequently, owing probably to the special wind pattern required for their existence. A more frequently occurring storm, which can also be

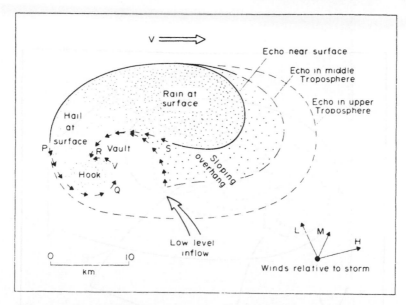

FIG. 13.5. Schematic diagram showing horizontal sections at three levels
through the radar echo of a supercell storm. (From Browning, 1964.)

large and severe, is the "multicell" storm (Fig. 13.7). Individual thunderstorm cells develop successively on the right hand side of a large storm
complex. Though each cell has a limited life cycle, the systematic
development of new cells produces a long-lived storm. The distinction
between supercell and multicell storms may not always be clear; some
storms exhibit a supercell shape yet on close inspection are found to
contain small-scale elements of short lifetime.

Vertical air velocities in thunderstorms can be measured with vertically
pointing Doppler radar, as in Fig. 13.2, or can be calculated from
measurements with two or more Doppler radars viewing the same storm
approximately horizontally from different directions. The raw data in
this approach consist of radial velocity components measured by radars
located at distances far enough from the storm to observe its full extent
at low elevation angles. First the field of the horizontal velocity components u and v is computed by combining the separate fields of radial
components. Then the vertical velocity w is obtained by integrating over
altitude the mass continuity equation,

$$\frac{\partial w}{\partial z} + \frac{w}{\varrho}\frac{\partial \varrho}{\partial z} = -\left(\frac{\partial u}{\partial x} + \frac{\partial v}{\partial y}\right),$$

in which the air density ϱ and its vertical variation are determined from a
temperature sounding. This procedure requires careful combining of the
radial velocity data and judicious smoothing of the computed fields, but

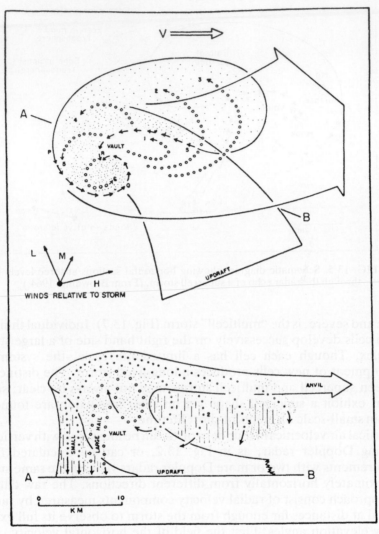

FIG. 13.6. Horizontal and vertical sections of airflow and precipitation trajectories in a supercell. (From Browning, 1964.)

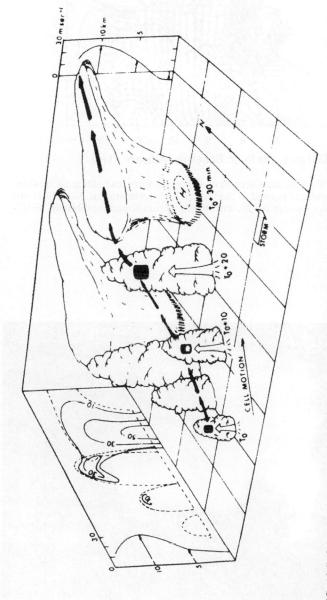

FIG. 13.7. Schematic view of a multicell storm. At the initial time the storm consists of four cells at different stages of development. The development of the youngest (southernmost) cell at successive times is indicated. The heavy dashed arrow is the trajectory of a parcel in the growing cell. A vertical section of the radar echo at the initial time is shown, as well as an indication of the wind profile. (From Chisholm and Renick, 1972.)

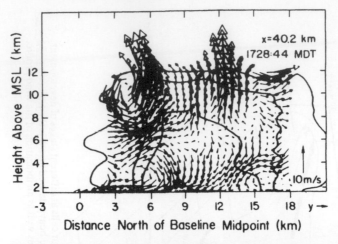

FIG. 13.8. Vertical cross section of a thunderstorm, indicating air motion, in a plane oriented normal to the direction of storm motion. Solid lines are reflectivity contours at 20, 28, and 36 dBz. (From Kropfli and Miller, 1976.)

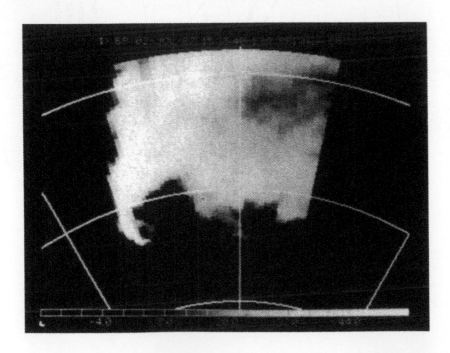

FIG. 13.9. A Colorado tornado (above) and its radar pattern (opposite).
The photograph was taken from a distance of a few kilometers. The black
funnel cloud has a diameter of about 300 m and consists mainly of soil blown
up from a freshly plowed field. The tornado was only 13 km from the radar,
affording an extremely high-resolution view. Range rings are at 5-km
intervals. The enclosed echo at the end of the hook-shaped appendage
corresponds exactly to the position of the tornado. The maximum Doppler
velocities observed in this echo were about 35 m/s. (Photos courtesy of
Brooks Martner, Wave Propagation Laboratory, NOAA.)

has been refined to give patterns of vertical velocity in thunderstorms such as in Fig. 13.8. This shows a complex circulation within the storm, several large vortices with scales of a kilometer or larger, and a predominance of updrafts at high levels and downdrafts below. Such observations indicate the important connection that can exist between the precipitation-induced cold downdraft and the inflowing air—connections that previously could only be inferred indirectly.

One of the most destructive manifestations of a thunderstorm is the tornado, an example of which is shown in Fig. 13.9. Tornadoes form in supercell thunderstorms that have developed in environments of strong wind shear and large convective instability. Observations with Doppler radars have shown that a precursor of tornadoes is the development of a mesocyclone—a horizontal circulation about 10 km across with values of vertical vorticity in the order of 10^{-2} s^{-1}. The details of tornadogenesis are not completely understood, but evidently the circulation within small regions of the mesocyclone can become intensified by convergence and the conservation of angular momentum to produce intense vortices with a horizontal extent of a few hundred meters. For a review of the theory of tornadoes the reader may consult Rotunno (1986).

Precipitation production by thunderstorms

Water that condenses in a thunderstorm updraft is either present in the form of cloud or precipitation or has evaporated. The mass $\mathfrak{M}$ that has condensed up to time t may be written

$$\mathfrak{M}(t) = C(t) + P(t) + F(t) + E(t),$$

where C is the mass of cloud water at time t, P is the mass of precipitation aloft, F is the mass of precipitation that has reached the ground, and E is the amount of cloud and falling precipitation that has evaporated. From a series of radar observations of the three-dimensional structure of a thunderstorm, it is possible to obtain estimates of two of the terms in this water budget. Using an empirical relation between the reflectivity factor and the precipitation content (a $Z - L$ relation) and integrating over the storm volume provides an estimate of $P(t)$. Also, applying an empirical relation between reflectivity and rain rate (a $Z - R$ relation) to the data in a horizontal plane at a low altitude, and integrating over area, yields an estimate of the instantaneous outflow of precipitation, dF/dt. Integrating this outflow up to time t gives $F(t)$.

The rate of generation of precipitation, g, may be defined as the rate of accumulation aloft plus the rate of outflow. That is,

$$g(t) = \frac{dP}{dt} + \frac{dF}{dt}.$$

The cumulative amount generated up to time t is then given by

$$G(t) = \int_0^t g(s)ds = P(t) + F(t).$$

The ratio $\tau = P/(dF/dt)$ has dimensions of time and may be thought of as the characteristic time of the precipitation process. All these quantities can be evaluated from radar data. Several studies of this kind, summarized by Rogers and Sakellariou (1986), led to the following conclusions:

1. The precipitation content P of a typical isolated thunderstorm is about 1 Tg (10^9 kg) during its mature stage of development.
2. The rate of outflow dF/dt is about 1 Gg/s (10^6 kg/s).
3. P and dF/dt fluctuate during a storm's history, but are closely correlated with each other.
4. The characteristic time of the precipitation process averages about 20 min during the mature stage.
5. The total outflow from a thunderstorm during its lifetime can exceed the amount P that is present in the cloud at any time during its mature stage by a factor of five or greater. Equivalently, the lifetime of a thunderstorm can exceed the characteristic time by a factor of five or greater.

Very large thunderstorms or mesoscale storm-complexes can have precipitation contents and outflow rates far in excess of these estimates.

Hail growth

Hailstones are formed when either graupel particles or large frozen raindrops grow by accreting supercooled cloud droplets. Thunderstorms contain both graupel and large drops, and it is not known which serves most frequently as the hail "embryo", although photographic evidence generally points to graupel. An important aspect of hail growth is the latent heat of fusion released when the accreted water freezes. Owing to this heating, the temperature of a growing hailstone is several degrees warmer than its cloud environment. In the theory of hail development, the temperature is determined by assuming a balance condition for the hailstone heating rate.

The heating rate due to the accretion of supercooled liquid droplets is given by

$$\frac{dQ_L}{dt} = \pi R^2 EMu(R)[L_f - c(T_s - T)], \tag{13.1}$$

where R is the radius of the stone and $u(R)$ its fall speed, L_f is the latent

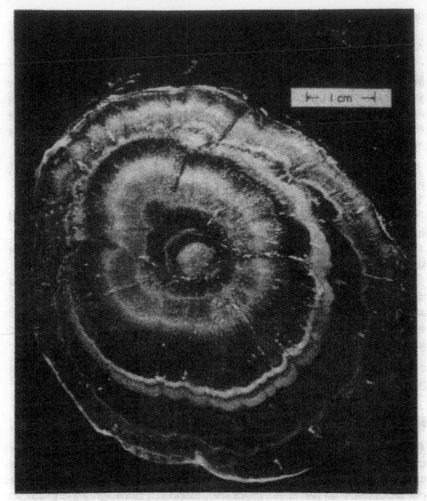

FIG. 13.10. Cross section of a large hailstone, showing the characteristic layered or "onionskin" structure. (Courtesy B. L. Barge, Alberta Research Council, Edmonton, Can.)

heat of fusion, c the specific heat of water, M the cloud liquid water content and E the effective collection efficiency, and T_s and T the temperatures of the hailstone and the ambient cloud.

The heat gained by sublimation is

$$\frac{dQ_v}{dt} = 4\pi RD(\varrho_v - \varrho_{vR})L_s a, \qquad (13.2)$$

where ϱ_v and ϱ_{vR} denote the ambient vapor density and that at the surface of the stone, L_s is the latent heat of sublimation, and a is a ventilation factor depending on hailstone size.

The rate at which heat is lost to the air by conduction is

$$\frac{dQ_s}{dt} = 4\pi RK(T_s - T)b,$$ (13.3)

where K is the heat conduction coefficient of air and b is a ventilation factor.

Equilibrium exists when

$$\frac{dQ_L}{dt} + \frac{dQ_v}{dt} = \frac{dQ_s}{dt},$$ (13.4)

which may be used to solve for the hailstone equilibrium temperature as a function of size, for given cloud conditions.

The rate of hailstone growth may be determined to a good approximation by adding the separate rates of growth by accretion and by sublimation. Accretion is usually dominant, and becomes more so as the stone grows. If it remains in the supercooled cloud long enough, the stone reaches the size for which the equilibrium temperature is 0°C, because of insufficient heat transfer to the surrounding air. (Typically this might occur for a diameter of about 1 cm.)

When the surface of the hailstone is at a subfreezing temperature, the collected water droplets freeze quickly and the surface remains essentially dry. When the surface is at 0°C, however, the collected water does not freeze immediately and the surface is wet. Although some water may be shed by the warm stone, much can remain to be incorporated into the stone, forming what is called "spongy ice". It has been deduced that the liquid fractions of large hailstones may amount to 20% or more. The entrapped liquid can later freeze if the stone enters colder or less dense cloud, where the heat transfer will suffice to chill the stone below 0°C. During its lifetime, a stone may undergo alternate wet and dry growth as it passes through a cloud of varying temperature and liquid water content, thus developing the layered structure that is often observed (Fig. 13.10).

Problems

13.1. During the time of its maximum development, a cumulonimbus cloud covers a horizontal area of 200 km^2 and produces rain at the ground at a rate of 2 Tg/h.

   (a) What is the area-average rain rate in mm/h?
   (b) Estimate the rate of release of latent heat by this storm, assuming that the condensation rate equals the precipitation rate.
   (c) Assuming that the updraft is continuous in height and occupies 20% of the storm volume, and that the difference in saturation mixing ratio between cloud base and cloud top is 8 g/kg, estimate the vertical air velocity required to produce precipitation at the rate observed.
   (d) If some fraction of the rain evaporates before reaching the ground, how will this affect the answers to (b) and (c)?

13.2. Crop damage caused by hail is approximately proportional to the kinetic energy imparted by the stones to the ground per unit area and time. Suppose there are N hailstones per unit volume of air, all with diameter D. Assume that their fall velocity is related to diameter by

$$u(D) = kD^{1/2},$$

where $k = 1.4 \times 10^3 \text{ cm}^{1/2} \text{ s}^{-1}$. Solve for the kinetic energy flux, F_{KE}, and the mass flux, F_M, of the hail at the ground. Show that they are related by

$$\frac{F_{KE}}{F_M} = \frac{k^2}{2} D.$$

13.3. Rain also imparts kinetic energy to the ground, though it is partly dissipated in overcoming surface tension when the drops break apart. Suppose that a population of raindrops has the general exponential form,

$$N(D) = N_0 \exp(-aD),$$

where N_0 is a constant, a is a parameter that depends on rainfall rate, and $N(D)dD$ is the number of drops per unit volume whose diameters are in dD. Suppose further that the fall velocity of the raindrops is given by the same dependence on diameter as in problem 13.2. (This is a reasonable approximation for large drops.) Solve for the kinetic energy flux, f_{ke}, and the mass flux, f_m, imparted to the ground by the raindrops, in terms of the quantities N_0, k, and a. Show that

$$\frac{f_{ke}}{f_m} = 2.25 \frac{k^2}{a}.$$

Suppose the rainfall rate is 20 mm/h and solve for the kinetic energy flux in units of W/m^2, assuming that $N_0 = 0.08 \text{ cm}^{-4}$.

13.4. Consider again the hailstones in problem 13.2. Suppose they are all of 1-cm diameter and that they are falling with a mass flux equivalent to that in a rainfall rate of 20 mm/h. Show that the kinetic energy flux of these hailstones is five times the kinetic energy flux of the equivalent rainfall.

13.5. A hailstone with a diameter of 2 mm begins to fall from a height of 5 km above cloud base, where the ambient temperature is 250 K. It grows by accretion of cloud water under conditions such that its surface temperature is constant at 0°C. The air has a lapse rate of 6°C/km within the cloud. Assume zero updraft velocity; assume also that during growth a balance always exists between the rate of heat gained by the freezing of accreted water and the rate of heat loss by conduction to the air. Neglect sublimation effects and the heat capacity of the collected water. Show that the hailstone grows to a diameter of approximately 7 mm after falling 3 km. Assume that the fall speed depends upon diameter as given in problem 13.2. For the ventilation factor, use $b = 0.3(Re)^{1/2}$, where Re is the Reynolds number of the flow about the stone.

13.6. A hailstone at 700 mb is falling through a cloud of supercooled droplets having a liquid water content of 4 g/m^3 and a temperature of −15°C. Determine the dependence of hailstone temperature on size, assuming equilibrium growth conditions, over the temperature range from −15°C to 0°C. Take into account heat transfer by conduction, sublimation, and accretion. Use the same fall speed and ventilation approximations as in problem 13.5, and assume that the ventilation factors a and b are equal.

13.7. When falling through thin cloud or cloud-free air a hailstone is likely to be cooler than its environment because of sublimation from its surface and also because of its heat capacity, which causes a time lag in its response to environmental temperature changes. Derive an expression for the thermal time constant of a hailstone, assuming that the temperature structure within the stone at any time is isothermal and that heat is transferred only by conduction to the air. That is, in the hailstone heat transfer equation, neglect all but the conduction term. Evaluate the time constant of a hailstone of 1-cm diameter, taking the specific heat of the stone to be 0.5 cal g^{-1} K^{-1} and assuming a ventilation factor of 25.

14

Weather Modification

IN THE late 1940s Irving Langmuir and his colleagues at the General Electric Research Laboratory in Schenectady, New York, discovered that dry ice, when dropped into a supercooled cloud deck from an airplane, caused a rapid conversion of the water to ice, leading quickly to the production of snowflakes and the dissipation of cloud in the region seeded. They had found a way to trigger an instability and convert cloud to precipitation. Soon thereafter they discovered that silver iodide had the same nucleating ability. These remarkable events, fondly described by Havens, Jiusto, and Vonnegut (1978), rejuvenated cloud physics. It is fair to say that much of the interest in the physics of clouds over the past few decades has been motivated, directly or indirectly, by the hope that improved understanding would lead to methods for predictable and reliable weather modification through cloud seeding.

Experiments in cloud modification are usually undertaken with one of the following goals: (1) to stimulate precipitation; (2) to dissipate cloud or fog; (3) to suppress hail. Widely varying degrees of success have been reported for each type of experiment. A fundamental problem is to distinguish the effects of seeding from natural variations. Except for smooth overcast clouds of the kind seeded in the earliest experiment, clouds are highly variable in structure and behavior, so that seeding effects can easily be masked by or confused with natural processes.

Because the potential economic benefits of even modest changes in precipitation are so great, cloud seeding was quickly appropriated as an activity for commercial exploitation. Zealous entrepreneurs often promised more than they could deliver. Economic pressures have made it difficult to sustain experimental programs in cloud seeding which, unlike operational seeding, do not hold the promise of immediate reward. Adding to the confusion, experimental seeding has sometimes been conducted in a rushed, haphazard way, without the proper controls to allow an objective assessment of results.

Although the prospects for substantial weather modification by cloud seeding appear dimmer now than a decade ago, evidence has emerged from a large body of research and a few carefully conducted experiments that modest changes in cloud structure and precipitation can be effected by seeding with suitable material in appropriate conditions. This chapter outlines some of the underlying principles.

A controversial subject, fraught with uncertainties, cloud seeding has received a well balanced scientific assessment by Dennis (1980), whose book may be consulted for information on the technology of cloud seeding and for a survey of a large number of seeding experiments.

Stimulation of rain and snow

In principle, precipitation may be encouraged by exploiting one of the instabilities in a cloud system. Only in this way is it reasonable to expect a large result from a relatively small expenditure of effort or material—or in other words to have an efficient modification technique. Before they produce precipitation, all clouds are colloidally unstable: the droplets have the potential of being swept out by precipitation; only lacking are a few large drops that can grow by accretion of the droplets. A typical 1-mm raindrop is the result of 10^5–10^6 droplet captures. The development of rain in warm clouds requires the natural appearance of about one droplet in 10^5, or 1000 per cubic meter, with a radius of at least 20 μm, which is a size large enough to grow by collisions and coalescence. One approach to stimulating rain in warm clouds is therefore to introduce approximately this concentration of large droplets into the ascending air above cloud base. These drops stand some chance of ascending and sweeping out cloud droplets throughout their upward and downward trajectories. They must be large enough initially to be able to grow, but not so large that they fall out before spending much time in the cloud. Also, more water is required for seeding at the appropriate concentration if large drops are used. A suitable compromise is probably droplets in the radius interval 20–30 μm. Experiments in water-seeding have been attempted by spraying water from airplanes flying at cloud base, relying on diffusion to disperse the drops throughout the updraft region. These experiments have occasionally suggested modest precipitation enhancement, but have usually been inconclusive.

An alternative method for stimulating coalescence is to inject salt particles around cloud base to provide centers on which large cloud droplets (raindrop embryos) can form. Large particles are preferred, because they will give solution droplets with relatively large critical radii. From (6.7) it follows that the critical radius r^* increases with salt particle radius according to $r^* \propto r^{3/2}$. The natural background of large salt particles (probably sea salt) is about 0.1 per liter. Therefore salt seeding should be designed to equal this amount at least, and to exceed it by perhaps an order of magnitude. Salt seeding is more efficient than water seeding in the sense that a smaller mass of seeding material is needed to produce the same number of raindrop embryos. However, there are practical problems connected with using salt, such as the clumping of particles in humid conditions and the corrosion of equipment, which do not exist for water. There are also theoretical uncertainties about the time required for salt particles of various sizes to grow to their critical droplet radii. Only rather recently have there been any encouraging results from salt seeding.

The phase instability of supercooled water droplets in subfreezing clouds is exploited by seeding with ice-forming (glaciogenic) material. Once ice crystals form they will tend to grow rapidly by diffusion at the expense of the droplets. About one ice crystal per liter in the upper parts of a supercooled cloud is enough to lead to precipitation development. For temperatures of about $-20°C$ and colder this number will form due to natural nuclei. Introducing ice crystals artificially into warmer clouds, in concentrations of about 1 per liter, would be expected to promote precipitation. This technique would be relatively ineffective at temperatures warmer than about $-5°C$ because the ice crystal process is slow (see Fig. 9.4, p. 161).

The principal seeding agents used to form ice crystals are dry ice (solid CO_2) and silver iodide (AgI). The equilibrium temperature of subliming CO_2 is $-78°C$, considerably colder than even the homogeneous freezing temperature of water. Injected into a supercooled cloud, pebble-size pieces of dry ice descend and leave a trail of vast numbers of ice crystals. The sudden chilling of the air produces high supersaturations, activating cloud condensation nuclei and also creating embryonic water droplets by homogeneous nucleation. The droplets freeze as soon as they are formed, producing as many as 10^{15} ice crystals per kilogram of dry ice. Laboratory experiments have shown that AgI particles can be introduced to clouds in the form of smoke produced by burning certain compounds of silver. Though their effectiveness is impaired by exposure to sunlight and possibly by wetting at warm temperatures, the AgI particles are known from observation to be an efficient source of ice crystals, at least in some conditions.

The possible effects of artificially produced ice crystals on a cloud depend on whether it is of cumuliform or stratiform type. In the special case of a stratiform cloud whose top is supercooled but does not extend beyond about the $-15°C$ level, the natural precipitation process may proceed very slowly owing to the scarcity of ice nuclei at relatively warm temperatures. In such clouds the introduction of ice crystals near cloud top by seeding with AgI or dry ice may cause precipitation that would not otherwise occur. The amount of precipitation would be small because of the low liquid water contents of cold stratiform clouds. If the cloud is persistent, however, due to large-scale meteorological factors, it would be possible by repeated seeding to build up a significant accumulation of precipitation. In stratiform clouds with tops colder than $-15°C$ seeding would not be required to initiate precipitation, but in some cases might be used to alter slightly the location where rain or snow falls—for example on a watershed instead of a few miles downwind.

For relatively small and short-lived cumuliform clouds the effect of seeding would be to initiate freezing slightly earlier and at a lower altitude than where it would occur naturally. It is not obvious that this would have much subsequent effect on cloud development except in very special circumstances. At most, seeding might cause a small amount of rain or snow to fall from a cloud that would otherwise fail to precipitate, or cause the rain to fall a little sooner than it would without seeding. Large and long-lasting cumulonimbus clouds produce precipitation naturally; seeding with ice nuclei would not be expected to affect the amount of precipitation produced.

Associated with freezing is the release of latent heat of fusion, representing what may be termed the latent instability inherent in supercooled clouds. This instability can be triggered by seeding with freezing nuclei or dry ice. It will contribute to the buoyancy of the updraft and may be of crucial importance in some clouds. Under some conditions cumulus clouds are limited in vertical development because of trapping by a thin inversion layer aloft. The extra latent heat liberated by seeding might be just sufficient, in a small class of such cases, to enable the cloud to penetrate the inversion and extend much further.

Cloud dissipation

Low-cloud overcasts and fogs pose hazards around airports. The concept of dissipating clouds by seeding is much like that of precipitation enhancement. Large particles or ice nuclei are introduced to sweep out the cloud droplets, thus clearing an area temporarily. Silver iodide and dry ice have been used with some success for clearing supercooled fogs and overcasts. Warm fogs have proved more difficult to affect. Experiments using salt and sprays have been attempted since as early as 1938 (Houghton and Radford), but there are no operational systems employing this approach. Ice fog, a winter phenomenon in some northern localities, also remains an unsolved problem.

Hail suppression

Two arguments have been advanced for alleviating hail by cloud seeding with ice nuclei. The first involves freezing essentially all of the supercooled droplets in the upper parts of a potentially hail-producing cumulonimbus. This in effect kills the accretional growth process, eliminating the possibility of large hail formation. Although the nucleating efficiency of AgI is presumably high, with estimates typically about

10^{14} nuclei per gram of AgI at $-20°C$, the amount of material required to glaciate a cloud is excessive and much beyond the capability of any seeding system in current use.

The second argument is more modest in the requirement of seeding material and involves adding ice nuclei only to the limited cloud region where hail is thought to have its maximum growth rate. Soviet scientists assume that this is the region in the upper part of the cloud where maxima in radar reflectivity are occasionally observed. They seed this region with AgI-charged artillery shells and report spectacular success in eliminating hail. The reason for the apparent effectiveness of this technique is not entirely clear.

A variation of the Soviet approach is to add the ice nuclei in a region lower down which is presumed to be the main updraft area. This region contains the natural ice nuclei or precipitation particles that are hail embryos. It is argued that by introducing artificial nuclei it may be possible to cause enough competition for the available water supply to make it unlikely that any hailstone will grow to a large size. In short, this approach is intended to create a large number of small stones instead of a few large ones. The small stones stand a chance of melting completely before reaching ground, or at least of causing less damage than the large ones.

Problems

14.1. One of the proposals put forward for suppressing hail is to seed the storm with ice nuclei in the updraft region where the natural ice crystals that develop into hailstones originate. This idea assumes that the artificial hail "embryos" will grow in the ascending air as graupel particles along with the natural embryos, leading to a high concentration of small hailstones in the cloud region of high liquid water content, where accretional growth would ordinarily be most rapid. These small hailstones compete for the available liquid water, and if their concentration is high enough none will be able to grow to the size of large, damaging hail. The critical question in this scheme is how many artificial embryos to introduce.

One way to approach the problem is to compare at a given altitude the rate at which liquid water is being made available by condensation in the updraft with the rate at which this water is used up by accretion. When the concentration of small hailstones exceeds some critical value, the depletion rate will exceed the production rate. This may be looked upon as the *minimum* hailstone concentration for hail suppression to be effective.

Set up the equations needed to analyze the problem from this point of view and estimate the minimum required hailstone concentration under the following assumptions:

(a) all hailstones are the same size, spherical, with diameter of 2 mm and terminal fall speed of 8 m/s,
(b) altitude about 6 km,
(c) updraft speed 25 m/s,
(d) growth by sublimation negligible,
(e) cloud liquid water content 6 g/m^3,
(f) collection efficiency of unity.

14.2. The supercooled portion of a convective cloud is rapidly converted to ice by seeding with a material that promotes freezing. At the 600 mb level the cloud temperature is initially $-20°C$ and the liquid water content is 3 g/kg. Calculate the total increase in temperature at this level as a result of complete glaciation of the cloud liquid.

Saturation mixing ratios over ice and water at 600 mb are given for various temperatures in the following table:

$T(°C)$	w_s (g/kg)	w_i (g/kg)
−23	1.002	0.800
−22	1.094	0.883
−21	1.194	0.973
−20	1.303	1.072
−19	1.420	1.179
−18	1.546	1.296
−17	1.682	1.425
−16	1.824	1.565

14.3. Inserting dry ice into a supercooled cloud promotes the production of ice crystals by local chilling, which may cause existing droplets to freeze and which may activate cloud condensation nuclei or cause homogeneous nucleation of droplets by creating high values of supersaturation. Suppose the cloud is initially saturated with respect to water at temperature T. Show that if the air is chilled to a temperature of $-40°C$, where homogeneous freezing is expected to occur, the supersaturation ratio will exceed 5.0, making homogeneous nucleation of water droplets likely, if $T > 250$ K.

15

Numerical Cloud Models

IN THE preceding chapters, we discussed the microphysical processes that govern the development of precipitation, and pointed out that these processes are closely coupled with the air circulation. When a cloud first forms, the concentration, sizes, and thermodynamic phase of the hydrometeors are determined by the air motions in and around the clouds, together with the characteristics of the aerosol particles that serve as condensation and freezing nuclei. Subsequent broadening of the size spectrum to produce precipitation is the result of collision, coalescence, accretion, riming, and aggregation, but these in turn are influenced by the airflow, especially the strength and duration of the updraft and the extent of mixing between the cloud and its environment. The veering and shearing of the environmental wind often determines the type of convection that can develop, and then exerts an influence on the spatial distribution of the precipitation. On the other hand, the microphysical processes of condensation, freezing, melting, and evaporation produce heat sources and sinks, which strongly affect the air circulation. The release of latent heat increases the buoyancy while the drag force of the falling particles causes the opposite effect. Thus there are complex feedbacks between microphysics and dynamics throughout the life of a cloud.

To gain a better understanding of precipitation mechanisms, the microphysical and dynamical processes should be examined as a coupled system. Because of the complexity of the cloud physical processes and the highly nonlinear nature of air motions, analytic solutions to problems in cloudy convection are extremely rare. The great range of scale in precipitation systems, from micrometers to kilometers and beyond, makes laboratory studies difficult if not impossible. The alternative is numerical simulation, which has emerged as a useful tool in the investigation of the interactions between microphysics and dynamics and between cloudy convection and the larger-scale environment. Research in the numerical simulation of clouds is advancing rapidly, and in the past decade highly sophisticated convective cloud models have been developed that have yielded new insights into storm processes.

The purpose of this chapter is to outline the fundamentals of cloud modeling and to demonstrate the usefulness of this approach with a few results. Our emphasis will be on physical insights; numerical and computational details, though an important part of the subject, are not included. For this information the reader must turn to the references. As a convenience, the models are classified according to the number of spatial dimensions that are included in the simulations.

The governing equations

Three kinds of processes must be accounted for in cloud models: dynamic, thermodynamic, and cloud physical. The equations we require are Newton's law of motion applied to the air, the continuity equation for air, an equation for temperature, and conservation equations for water in its different forms. Because cloudy convection can be regarded as a perturbation about an ambient environment that is in hydrostatic equilibrium, the equations can be written to describe the deviations of the different variables from the ambient basic state.

The pressure, density, and temperature are written as the sum of a basic state variable (p_0, ϱ_0, T_0) and a perturbation ($\hat{p}$, $\hat{\varrho}$, $\hat{T}$) from the basic state:

$$\left. \begin{array}{l} p = p_0(z) + \hat{p} \\ \varrho = \varrho_0(z) + \hat{\varrho} \\ T = T_0(z) + \hat{T} \end{array} \right\} . \tag{15.1}$$

If the basic state is in hydrostatic equilibrium and all perturbation quantities are small, the force on the parcel in the vertical direction can be written as

$$-\frac{1}{\varrho}\frac{\partial p}{\partial z} - g = -\frac{1}{\varrho_0}\frac{1}{\left(1 + \dfrac{\hat{\varrho}}{\varrho_0}\right)}\frac{\partial}{\partial z}(p_0 + \hat{p}) - g = -\frac{1}{\varrho_0}\frac{\partial \hat{p}}{\partial z} - g\frac{\hat{\varrho}}{\varrho_0} . \tag{15.2}$$

This equation, unlike elementary parcel theory, does not require the pressure of the parcel to adjust instantly to that of the ambient air.

Applying the perturbation expansion to the ideal gas law leads to

$$\frac{\hat{p}}{p_0} = \frac{\hat{T}}{T_0} + \frac{\hat{\varrho}}{\varrho_0}, \tag{15.3}$$

which, when substituted into (15.2), gives

$$-\frac{1}{\varrho}\frac{\partial p}{\partial z} - g = -\frac{1}{\varrho_0}\frac{\partial \hat{p}}{\partial z} + g\frac{\hat{T}}{T_0} - g\frac{\hat{p}}{p_0} . \tag{15.4}$$

This equation reduces to (3.9) if the perturbation pressure is neglected.

To allow for the effect of water vapor on buoyancy, the equation can be generalized by replacing temperature with the virtual temperature.

To describe convection with precipitation in two dimensions requires at least seven differential equations: one each for the horizontal and vertical components of air velocity, the continuity equation, the equation for temperature, and one each for water substance in the form of vapor, cloud, and precipitation. The horizontal and vertical velocity components will be denoted by u_1 and u_3 respectively, the virtual temperature by T_v, and the mixing ratios of vapor, cloud, and rain, by w, μ, and R. The equations are then as follows:

$$\frac{\partial u_1}{\partial t} = -u_1 \frac{\partial u_1}{\partial x} - u_3 \frac{\partial u_1}{\partial z} - \frac{1}{\varrho_0} \frac{\partial \hat{p}}{\partial x} + F_{u_1}, \tag{15.5}$$

$$\frac{\partial u_3}{\partial t} = -u_1 \frac{\partial u_3}{\partial x} - u_3 \frac{\partial u_3}{\partial z} - \frac{1}{\varrho_0} \frac{\partial \hat{p}}{\partial z} + g\left(\frac{\hat{T}_v}{T_{v_0}} - \frac{\hat{p}}{p_0} - q\right) + F_{u_3}, \tag{15.6}$$

$$0 = \frac{\partial}{\partial x}(\varrho_0 u_1) + \frac{\partial}{\partial z}(\varrho_0 u_3), \tag{15.7}$$

$$\frac{\partial T}{\partial t} = -u_1 \frac{\partial T}{\partial x} - u_3\left(\frac{\partial T}{\partial z} + \Gamma\right) + F_T + \phi_T, \tag{15.8}$$

$$\frac{\partial w}{\partial t} = -u_1 \frac{\partial w}{\partial x} - u_3 \frac{\partial w}{\partial z} + F_w + \phi_w, \tag{15.9}$$

$$\frac{\partial \mu}{\partial t} = -u_1 \frac{\partial \mu}{\partial x} - u_3 \frac{\partial \mu}{\partial z} + F_\mu + \phi_\mu, \tag{15.10}$$

$$\frac{\partial R}{\partial t} = -u_1 \frac{\partial R}{\partial x} - u_3 \frac{\partial R}{\partial z} - \frac{1}{\varrho_0} \frac{\partial}{\partial z}(\varrho_0 RV) + F_R + \phi_R. \tag{15.11}$$

In (15.6) q stands for $\mu + R$, the mixing ratio of the condensed water. The F_i terms represent changes caused by turbulent fluxes, and the ϕ_i terms are sources and sinks. In (15.11) the third term on the right describes the vertical divergence of rainwater whose effective fall velocity is V. The ice phase can be included by adding another conservation equation similar to (15.10) and (15.11). In the continuity equation (15.7) the local rate of change of density has been set to zero. This is called the anelastic assumption.

This set of equations is similar to the system derived by Ogura and Phillips (1962). Because of the anelastic constraint, it has the advantage that sound waves, which move at a fast speed, are not among the possible solutions. Relatively long time steps can then be used in the integrations without causing numerical instability. Although (15.5)–(15.11) constitute seven equations in the seven unknowns u_1, u_3, T, w, μ, R, and $\hat{p}$, the procedure is to use (15.7) to eliminate the time derivative terms in (15.5) and (15.6) and obtain a new equation for $\hat{p}$ which is solved instead of (15.7). Some researchers have returned to the fully compressible equations, which support sound waves and therefore require much finer time steps in the integrations. For example, Klemp and Wilhelmson (1978a) developed a thunderstorm model using the compressible system of equations. It employs a time-splitting numerical technique to alleviate the heavy computational requirements.

Present practice is to treat the F_i terms like diffusion, using an eddy diffusion coefficient that is a nonlinear function of shear and static stability or of turbulent kinetic energy (see for example Klemp and Wilhelmson, 1978a, and Yau and Michaud, 1982).

To complete the model formulation, initial and boundary conditions also have to be prescribed. These usually consist of an observed or a hypothetical environmental sounding and a specification of surface heat and moisture fluxes. An important technicality is the treatment of lateral boundary conditions. Edge effects caused by a limited computational domain can give spurious results unless the boundary conditions are properly formulated. The so-called radiation boundary condition, which allows disturbances to propagate out of the domain, is a favored choice. The reader is referred to Hedley and Yau (1988) for a discussion of this topic.

One-dimensional models

The first attempt at simulating the interactions between cloud dynamics and microphysics was through one-dimensional models. Representative of this kind of model is the one developed by Srivastava (1967). The ice phase and turbulence are neglected. Water is present in the form of vapor, cloud water, and rainwater. The sizes of cloud drops and raindrops are not specified; however, cloud drops move directly with the air and raindrops fall relative to the air with a velocity that depends on the amount of rainwater per unit volume. Details of the condensation process are neglected: any excess of vapor over the equilibrium value is assumed to condense out in the form of cloud water. Similarly, any deficit below the equilibrium value is supplied by evaporation of the available condensed water. Coalescence is neglected: cloud water is assumed to convert spontaneously to rainwater once a specified threshold in cloud water content is exceeded. Because of the limitation to one dimension, cloud quantities may vary only in the vertical.

The governing equations, which are the one-dimensional version of (15.6), (15.8)–(15.11), are as follows, where we have replaced u_3 by U:

Vertical velocity:
$$\frac{\partial U}{\partial t} + U \frac{\partial U}{\partial z} = g \frac{\hat{T}}{T_0} - gq. \qquad (15.12)$$

The last term represents the reduction in buoyancy by the weight of the condensed water. The effect of water vapor on buoyancy is neglected. As in most one-dimensional models, the perturbation pressure terms are not included. (See Yau, 1979, for a discussion of the importance of these terms.)

Water vapor:
$$\frac{\partial w}{\partial t} + U \frac{\partial w}{\partial z} = \phi_w = E, \qquad (15.13)$$

where E is an evaporation term that can be positive or negative.

Cloud water:
$$\frac{\partial \mu}{\partial t} + U \frac{\partial \mu}{\partial z} = \phi_\mu = -E_1 - P, \qquad (15.14)$$

where E_1 is a term that describes cloud evaporation or condensation, and P describes the production of rain from cloud by spontaneous coalescence (called autoconversion) and by accretion.

Rainwater:
$$\frac{\partial R}{\partial t} + U \frac{\partial R}{\partial z} + \frac{1}{\varrho_0} \frac{\partial}{\partial z} (\varrho_0 RV) = \phi_R = -E_2 + P, \quad (15.15)$$

where E_2 is a term that describes the rate of evaporation of rain.

Temperature:
$$\frac{\partial T}{\partial t} + U\left(\frac{\partial T}{\partial z} + \Gamma\right) = -\frac{L}{c_p} E, \qquad (15.16)$$

where $E = E_1 + E_2$ is the evaporation factor also appearing in (15.13).

The factors E_1, E_2, V, and P were expressed by Kessler (1969) as functions of the other dependent variables in the system on the basis of empirical arguments about drop-size distributions, fall velocities, and the conversion of cloud to rain. Kessler's expressions vastly simplify calculations, making it unnecessary to account specifically for drops of different sizes. These parameterizations have been employed directly or in modified form in most of the simpler cloud models, including that of Srivastava.

Equations (15.12)–(15.16) were solved by Srivastava for different combinations of initial conditions and assumptions. Figure 15.1 shows results from a case with the following initial conditions: the initial profiles of updraft and cloud water are the steady state solutions of (15.12) and (15.14) with R set equal to zero; no rain is present initially; U and μ are initially zero at cloud base; U is zero at the surface and remains so. The ambient temperature sounding is representative of convective rain conditions in warm climates, with a cloud base temperature of 4°C and a lapse rate of 6.8°C/km. At all altitudes except the very top of the cloud, the updraft is a maximum initially, when no rain is present. During the first 10 min rain develops at the expense of cloud water through the entire cloud, with the maximum rain production in the upper part of the cloud,

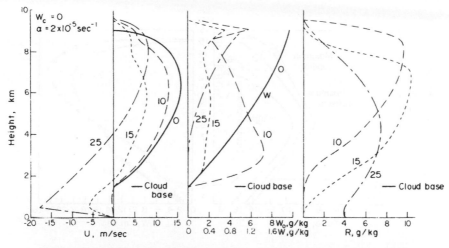

FIG. 15.1. Results of calculations for a one-dimensional cloud model, showing vertical profiles of updraft (left), cloud water mixing ratio (center), and rainwater mixing ratio (right) at the indicated times in minutes. Note that different scales are used for the cloud water content initially (W_0) and at subsequent times (W). (From Srivastava, 1967.)

where the cloud water content was initially a maximum. At 15 and 25 min the rain has descended, creating a low-level downdraft.

The interaction between water content and vertical velocity is illustrated in Fig. 15.2. At both altitudes depicted the updraft velocity weakens as the water content increases. After the water content reaches a peak, the updraft is allowed to increase, but this in turn increases the

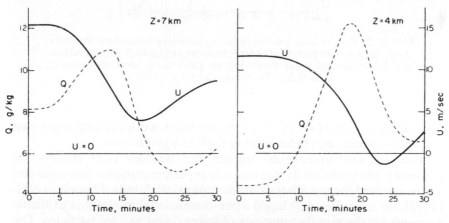

FIG. 15.2. Variation in time of updraft velocity U and total water content $R + \mu$ ($= Q$ in this figure) at heights of 7 km (left) and 4 km (right). (From Srivastava, 1967.)

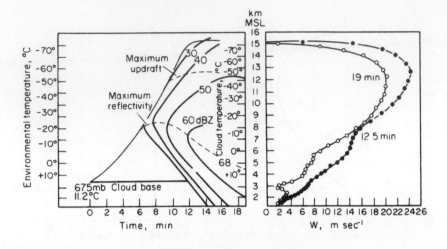

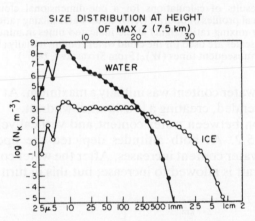

FIG. 15.3. Results from a model including extensive microphysical details; above, time–height pattern of radar reflectivity and updraft profiles; below, size distribution of water drops and ice particles at level of maximum reflectivity. (From Danielsen *et al.*, 1972.)

rate of production of water. Thus the feedback works in such a way that the water content lags behind the updraft at a given level.

Although one-dimensional models can account only crudely for dynamics, such models do not make heavy computational demands and are well suited for the analysis of microphysical details. Danielsen *et al.* (1972), for example, developed a one-dimensional model that explicitly accounts for the size distributions of water drops and ice particles. The size distribution of water drops is defined by 31 categories, arranged logarithmically from 2.5 μm to 2.5 mm in radius. Nine additional categories extend the ice particles to a radius of 20 mm. With the distributions thus described there is no need to use bulk parameterizations for autoconversion, condensation, and evaporation. The cloud-to-rain conversion can be determined more accurately than in simpler, parameterized models by incorporating stochastic collection. Breakup of large drops is modeled by forcing the size spectra toward the Marshall–Palmer distribution. An empirical method is also used to model freezing, such that freezing becomes possible as the temperature drops below $-7°C$, with large drops being preferentially frozen first, and all drops frozen for temperatures below $-45°C$. Entrainment effects are modeled by incorporating a lateral mixing term.

Figure 15.3 is an example of results that resemble the observations of an actual hailstorm. Danielsen *et al.* found that hail growth in the model depends on the updraft speed, the initial droplet distribution, the surface humidity, and the altitude at which the drops begin to freeze. Hail was favored in cases with a broad initial drop spectrum and in which the maximum updraft velocity is between 15 and 30 m/s. Maximum hail growth occurs in the mixed-phase region of the cloud, which is located beneath the updraft maximum.

Two-dimensional cloud models

One-dimensional models cannot include vertical shear of the ambient wind, which is often an important environmental factor in the development and behavior of clouds. Moreover, any simulation short of full three dimensions imposes unrealistic constraints on the air flow. But modeling in three dimensions also entails limitations because of the excessive computational requirements. To illustrate this point, we note that the simulation of a convective storm requires a computational domain of at least $50\,km \times 50\,km \times 15\,km$. To carry out the integrations with a spatial resolution of 100 m requires a total of about 4×10^7 grid points. Just to store the three-dimensional arrays for wind, temperature, pressure, humidity, cloud water, and rain water requires 3.2×10^8 words of memory, an amount that is several times the central core capacity of most computers. To resolve the fine-scale air motions within the cloud requires an even finer grid.

Because of computer limitations, researchers are often forced to assume either slab symmetry or axial symmetry, both of which reduce the three-dimensional problem to two dimensions, and greatly reduce the computational requirements. In axisymmetric models, the equations are formulated in cylindrical polar coordinates and the azimuthal gradients of all quantities are assumed to be zero. These models are only suitable for simulating an isolated, single-cell cloud in a calm environment, because the effects of ambient wind shear or the interactions between clouds cannot be incorporated. The equations in slab symmetric models are those set out in (15.5) to (15.11). These models can simulate conditions in a single vertical plane, assuming that the cloud properties do not vary in the direction orthogonal to the plane. Confining the flow to a vertical plane, however, does not allow for the realistic simulation of updrafts and downdrafts, and grossly overestimates the magnitude of perturbation pressure forces. Nevertheless, slab symmetric models are used widely because they allow for unidirectional shear in the ambient airflow and at the same time permit better spatial resolution than is possible in three-dimensional models.

Takeda (1971) was among the first to analyze the effects of wind shear on storm development using a two-dimensional numerical model. In his model the size distribution of water drops is described by seven discrete radii. Cloud and rainwater contents, rainfall intensity, and radar reflectivity are computed directly from the size distributions. The model includes condensation, evaporation, stochastic coalescence, spontaneous breakup, and the fall velocity of drops relative to the air. Diffusion is modeled by the inclusion of an eddy diffusion coefficient in the governing equations.

Convection is started by an initial disturbance of temperature, the extent and intensity of which are specified. Although the main interest was in the effects of ambient wind, Takeda also experimented with different values of the ambient stability and of the initial disturbance. He found that the clouds could be grouped into three types according to the way the cloud develops: type 1, in which new clouds develop around a dissipating cloud; type 2, in which the cloud is short-lived; and type 3, in which the cloud is long-lasting.

Type 1 clouds develop if the atmosphere is sufficiently unstable and the ambient wind shear is weak. A downdraft forms in the initial cloud because of precipitation loading and cooling by evaporation. The current of cold air from the downdraft spreads outward near the ground, causes lifting of the surrounding air, and stimulates the development of a new convective cloud.

If the wind is strongly sheared in the vertical, with the velocity monotonically increasing with height, the initial cloud is inclined in the same sense as the shear. The precipitation-induced downdraft forms on the downshear side of the cloud, and the convective upcurrent is tilted by the vertical wind shear. This arrangement intensifies the vertical shear on the downshear side of the cloud. Initiated by the cold air outflow, new upcurrents are damped by the intensified shear before they can develop a convective cloud circulation. The cloud is therefore short-lived.

When the wind profile consists of a jet or extremal value at lower levels, with layers above and below the jet having shear vectors in opposite directions, the patterns of rainwater content and updraft velocity are both inclined in the direction of the lower level shear. The rainwater is displaced from the updraft in the direction of the jet, leading to the formation of a downdraft. The dome of cold air originating from the downdraft creates a new upcurrent in the vicinity of the original updraft. In this situation the updraft and downdraft form an organized circulation that can last a long time. The convective cloud attains a steady state and is termed long-lasting. If the height of the jet is too low or too high it has little effect on the convection and a type-2 cloud develops.

Figure 15.4 illustrates the rainwater content and streamlines for a type-3 cloud in an environment with a low-level jet at a height of 2.5 km. By 30 min the precipitation-induced downdraft has been formed to the left of the upcurrent. This induces low-level upward motion to the left of the cloud, leading to the formation of a small raincloud at 45 min. The small cloud does not develop further, and at 60 min the structure of the parent cloud is similar to that at 30 min.

Takeda was one of the first researchers to include radar reflectivity in a cloud model. This is an important quantity to compute, because it is often the only means of comparing the structure of real clouds with model predictions. Takeda's calculated reflectivities agreed well in magnitude and time and space structure with those observed in large convective clouds.

Figure 15.5 gives another example of a two-dimensional simulation, in this case of a cumulus congestus cloud in its developing stage before precipitation has formed. The integration domain is 28 km in width and 9 km in depth. To maximize the resolution near the region where the cloud develops, the size of the horizontal grid interval decreases smoothly from 1.15 km near the boundaries to 87 m near the center of the domain. The vertical grid length is set constant at 90 m. Convection is initiated by an impulse of excess moisture near the level of the cloud base.

The figure shows the vertical cross section of the updraft velocity at 32.5 min after the start of convection for two simulations, designated as W1803 and X1803. In X1803 the ambient wind is set to zero and the updraft becomes concentrated in a narrow vertical channel. In W1803, a weak variation of wind with height is included. This causes the updraft to be tilted to the left in the lower part of the cloud and to the right above, consistent with the change in the direction of the wind shear. The maximum updraft velocity in W1803 is 15 m/s, a value significantly less than the maximum in X1803 (22 m/s). The wind shear hinders the "communication" between the upper and lower parts of the cloud core, suppressing the development of the cloud. Simulation W1803 also shows that the subsiding motion on the downshear side of the cloud (right side in the figure) is about four times stronger than on the upshear side, a result that agrees with observations of developing cumulus clouds.

To investigate the mechanism leading to the formation of the strong

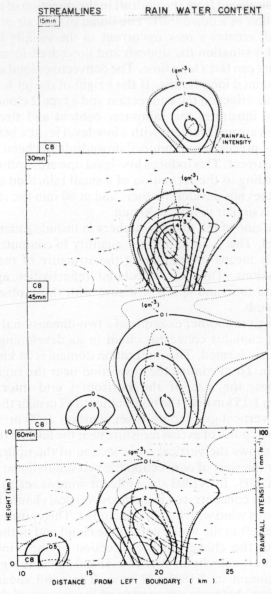

FIG. 15.4. Rainwater contents (g/m³) and streamlines in a simulated two-dimensional precipitating cloud growing in a sheared environment. The broken line indicates rainfall intensity at the 500 m level, and the dotted line shows the cloud boundary. (From Takeda, 1971.)

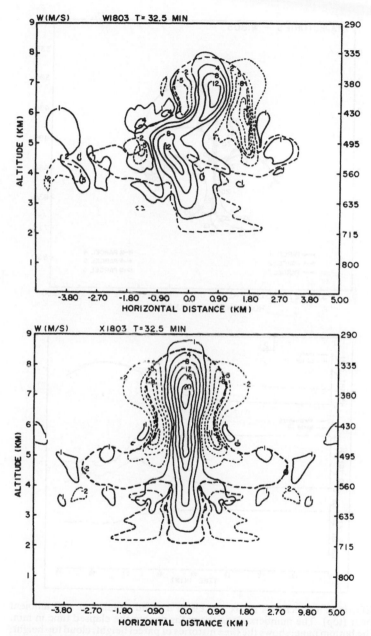

FIG. 15.5. Contour plots of vertical velocity for cloud simulations with ambient shear (W1803) and without shear (X1803) at 32.5 min. The thick dashed line denotes the 0.1 g/kg contour of cloud water mixing ratio. (From Reuter and Yau, 1987c.)

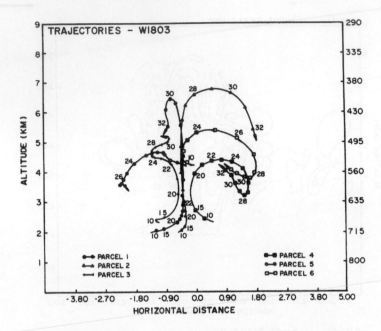

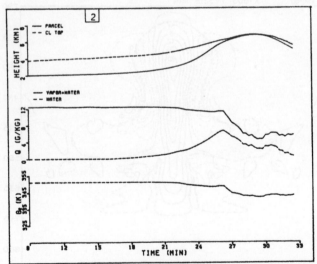

FIG. 15.6. Trajectories of six parcels in a cloud simulation with ambient shear (top). The numbers along the paths denote the elapsed time in min. The bottom panel shows the time histories of parcel height, cloud top height, total water mixing ratio Q, cloud water mixing ratio, and equivalent potential temperature θ_e for parcel 2. (From Reuter and Yau, 1987b.)

downdraft on the downshear side, Reuter and Yau followed the trajectories of 300 air parcels from different initial locations. Figure 15.6 illustrates the paths of six of these. Of special interest is parcel 2, which rises through the cloud to a level near the cloud top and then sinks into the region of strong downdraft on the downshear side. Its altitude, equivalent potential temperature θ_e, and total water mixing ratio $Q = w + \mu$, are plotted as functions of time. Because there is no precipitation in this simulation, Q and θ_e are conserved so long as mixing does not occur between the parcel and the ambient air. The results show that these quantities do remain relatively constant as the air parcel rises near the center of the updraft. But as the parcel approaches the cloud top at about 28 min, Q and θ_e decrease rapidly, indicating that mixing of cloudy and environmental air has occurred. Mixing causes evaporation, cooling, and a downward acceleration. A penetrative downdraft is formed, consistent with the cloud top mixing mechanism discussed in Chapters 4 and 5.

Using a slab symmetric model, Farley and Orville (1986) showed that many observed features of a severe hailstorm could be simulated realistically. In their model, cloud water, cloud ice, and rain are treated using bulk water parameterizations, while the graupel and hail are distributed over 20 logarithmically spaced size categories. The model reproduced the observed sloping updraft, domed top, recycled precipitation, radar overhang, and moving gust front, although it did not generate the large amount of hail that was observed. The model was also found useful in assessing the effects of glaciogenic cloud seeding.

Three-dimensional models

Clouds and storms usually develop in an environment in which the wind speed and direction change with height. Observations show that convective clouds interact with the ambient flow, acting as barriers at some levels, moving with the wind at others. They often exhibit rotational features, such as vortices or vortex pairs, which in extreme cases can be precursors to tornadoes. The outflow in downdrafts can interact with the ambient flow, leading to the formation of multicellular storms. Three-dimensionality is fundamental to these storms. To simulate them, three-dimensional cloud models have to be used.

The first fully three-dimensional cloud model was that of Steiner (1973). Comparing the development of a small, non-precipitating cumulus cloud for the same ambient conditions in a two-dimensional and a three-dimensional domain, he found important differences in cloud behavior. In two dimensions the vertical velocities were reduced and the magnitudes, and in some cases even the direction, of energy conversions were different. His work gave the first clear indication that while it is possible with simpler models to investigate precipitation development and some of the interactions between cloud dynamics and microphysics, three-dimensional models are needed to simulate the dynamics properly.

Significant progress has been made since then. Current three-dimensional models are able to simulate convincingly the large-scale characteristics of clouds, with particularly good results for severe storms in midlatitudes. Their use has led to a better understanding of the importance of mesoscale lifting in organizing and sustaining convective storms (e.g., Schlesinger, 1982) and to a clarification of details, such as the occasional splitting of storms (Wilhelmson and Klemp, 1978).

Radar observations show that a severe storm can sometimes split into two separate storms, one of which moves to the right and the other to the left of the ambient wind. Wilhelmson and Klemp succeeded in simulating this behavior, recognizing it as a challenging test for a numerical cloud model. The initial ambient conditions were idealized to allow the wind speed but not the direction to vary with height. At the surface, the wind is 8 m/s from the east. With no change in direction, the wind decreases approximately linearly with height, passes through zero, and becomes a west wind of 12 m/s at 4 km. Above 4 km the wind is constant at 12 m/s from the west. Thus the wind shear vector is approximately constant and from the west up to 4 km and is zero above that level. The wind hodograph, which is the curve defined by the end points of the wind vectors at successive altitudes, is simply a straight line in the east–west direction. Figure 15.7 shows some of the results. Because the Coriolis force was neglected and the initial conditions were symmetric about the vertical plane $y = 0$, only the results for the southern half of the storm are presented. The northern half is simply a mirror image of its southern counterpart.

The rainwater content at a height of 1.75 km shown in the left panel indicates three stages in the splitting. The formation stage at 30 and 45 min shows the precipitation intensifying and becoming elongated along the direction of the ambient wind shear. During the extension stage at 60 min, the rainwater spreads in a direction normal to the ambient wind. The splitting stage is shown at 75 min, when the center of the rain area has shifted from the axis of symmetry. The storms afterwards separate,

FIG. 15.7. Horizontal cross sections for model times 30, 45, 60, and 75 min showing the rainwater (1 g/kg contour interval) at 1.75 km (left), and the horizontal vector winds and the vertical velocity at 0.25 km (middle) and at 1.75 km (right). The maximum horizontal wind speed is indicated for each time in parentheses in m/s. The contour interval for vertical velocity is 0.5 m/s (middle) and 2 m/s (right). Solid contours are used in updrafts and dashed contours in downdrafts. The thick line is the outline of the rainwater region. (From Wilhelmson and Klemp, 1978.)

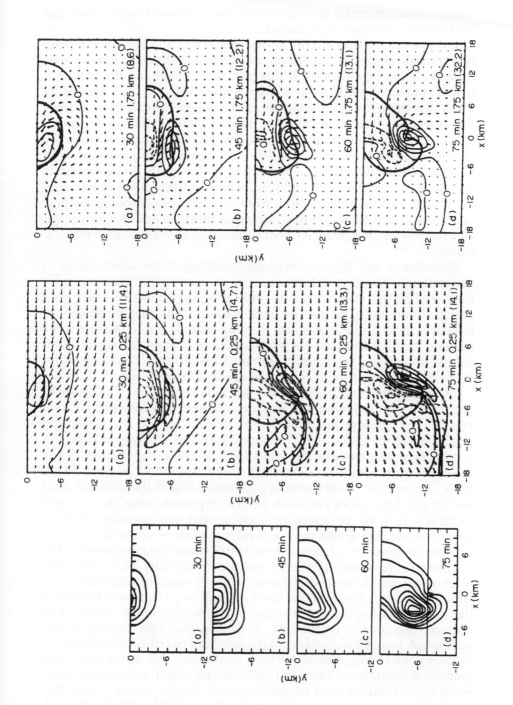

with the southern storm moving to the right of the wind shear vector and the northern twin moving to the left.

The low-level wind patterns in the middle panel of Fig. 15.7 explain how the splitting occurs. At 30 min, the storm is fed by moist low-level air from the east. Precipitation is reaching the surface and the low-level updraft begins to weaken. At 45 min, a downdraft has been formed by the drag of the rainwater and evaporative cooling. The diverging air from the downdraft interferes with the easterly flow to form a region of convergence south of the initial storm. The vertical air motion caused by the low-level convergence continues to strengthen and becomes oriented in a north–south direction by 75 min. The outflow from the downdraft is spreading toward the south and west. The gust front, a line delineating the abrupt shift in the horizontal wind, becomes noticeable in the new updraft region at 60 and 75 min. Moist air from the east rising over the cold outflow air along the gust front sustains the development of the new storm.

Wilhelmson and Klemp found that splitting is the direct result of precipitation. Splitting did not occur in simulations without precipitation, and if the evaporation of rain was suppressed in cases with precipitation, the split storms were weaker. The magnitude of the low-level shear also affects storm development. It was found that if the easterly flow is weak near the surface, the gust front propagates eastwards, away from the updraft region of the storms. The storms subsequently decay because their updrafts cannot be sustained.

The right panel in Fig. 15.7 shows the midlevel air flow. The southern split storm or the right-mover is associated with a cyclonically rotating vortex. The ambient air from the south streams into the downdraft region, descends, and spreads out near the surface causing the region of convergence that sustains the split storm. Because the strength of the low-level convergence is related to the midlevel inflow into the downdraft, Klemp and Wilhelmson hypothesized that if the ambient wind hodograph turns clockwise with height, especially near the low and midlevels, such that the southern inflow at midlevels is stronger than for a straight-line hodograph, the low-level convergence would be increased and the development of the southern storm would be strengthened. The northern storm would experience exactly the opposite effect. The midlevel inflow into the downdraft from the north would be inhibited by the stronger ambient flow from the south. Therefore, a clockwise turning of the hodograph implies the selective strengthening of the southern or right-moving storm. Conversely, a counterclockwise turning of the hodograph favors development of the northern or left-moving storm.

Klemp and Wilhelmson (1978b) confirmed this hypothesis with numerical simulations. They also explained why right-moving storms are much more common than left-moving ones, for the simple reason that the environmental wind usually turns clockwise with height. Klemp *et al.*

(1981) offered an alternate interpretation of the process of selective development in terms of pressure and vorticity distributions.

Using observed rather than idealized ambient winds, Wilhelmson and Klemp (1981) were able to reproduce many of the observed characteristics of a severe storm, such as the direction of movement, the timing of splitting, and the shape of the precipitation pattern. In fact, the agreement they found between simulations and observations was remarkable. A computer model that simulates nature so closely can provide detailed insights to dynamic and thermodynamic processes that are not available from radar or other observations.

Convective storms are sometimes observed to develop a pair of counterrotating, mesoscale vortices, typically 3–10 km across. Tornadoes are thought to develop by the intensification of a mesovortex. Three-dimensional cloud models have simulated mesovortices and their intensification. Figure 15.8 compares the observed and simulated low-level structure of a cyclonically rotating mesovortex. The close resemblance of the simulation and the actual storm makes it possible to investigate quantitatively the different processes responsible for the development of rotation and the transition of the storm to its tornadic phase.

Three-dimensional cloud models have succeeded in simulating multicellular midlatitude storms (e.g., Wilhelmson and Chen, 1982) and convection in the tropics (e.g., Turpeinen and Yau, 1981). With the help of increasingly powerful computers, models have also been able to simulate the development of an ensemble of cumulus clouds. Figure 15.9 shows a simulated group of cumulus clouds that were initiated by random heating at the surface. At the beginning, longitudinal roll clouds with their axes parallel to the direction of the mean wind are formed. By 106 min the longitudinal structure can still be recognized although some clouds are exhibiting a cellular structure. The clouds are found to interact strongly with one another, leading to the formation of new clouds by the convergence of downdrafts from neighboring clouds.

If detailed observations of the surface fluxes of heat and moisture are available, it is possible to initiate a field of cumulus clouds without the need of random heating. Figure 15.10 shows an example of a simulated cumulus ensemble that was initiated using a surface boundary layer model incorporating observed heat and moisture fluxes. It is found that realistic surface layer forcing leads to cloud ensembles that agree closely with observations. Sensitivity tests of the separate effects of topography and surface fluxes suggest that the global characteristics of the cloud field, such as the cloud size distribution and the fractional cloud coverage, are mainly determined by surface orography. On the other hand, thermodynamic inhomogeneities created by variations in soil type and vegetation influence the temporal and spatial location of an individual cloud in the cloud ensemble.

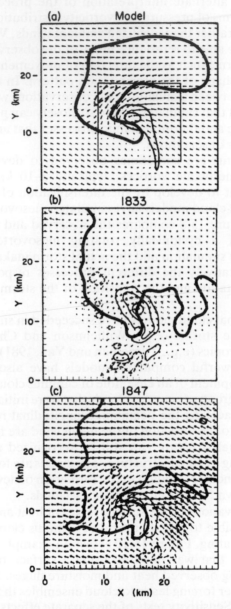

FIG. 15.8. Horizontal cross sections at height 1 km for model simulation (top) and observations (below) at 1833 and 1847. Updraft velocities (solid lines) and downdraft velocities (dashed lines) are contoured at 5 m/s for simulation at 1833 and at 10 m/s for 1847. The heavy solid line denotes the 0.5 g/kg rainwater contour in the model and the 30 dBz reflectivity contour in the observed storm. Horizontal wind vectors are scaled so that one grid interval represents 20 m/s. (From Klemp *et al.*, 1981.)

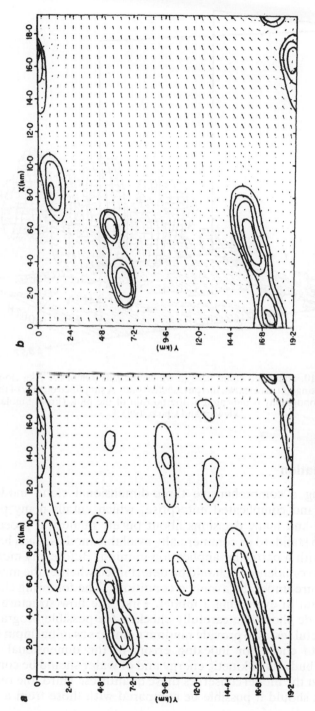

FIG. 15.9. Horizontal vector velocity and cloud water mixing ratio at (a) 1950 m and (b) 2550 m. The model time is 106 min and the contour interval is 0.5 g/kg. The scale of the longest arrow is 4 m/s. (From Yau and Michaud, 1982.)

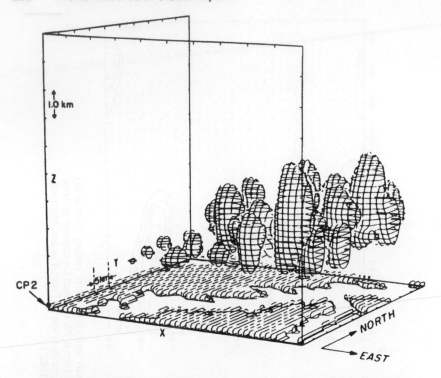

FIG. 15.10. Simulated cloud field in which the surfaces of 0.1 g/kg cloud water mixing ratio are shown in a perspective plot. A simplified display of the surface topography is indicated at the bottom of the plot. (From Smolar-kiewicz and Clark, 1985.)

Model evaluation

Developing a comprehensive three-dimensional cloud model is a challenging and complicated enterprise, which entails many physical assumptions and many decisions about purely numerical procedures. There are no standard or commercially available models; each has been developed with a particular application in mind. Before any model can be used with confidence, it should be determined whether the model can simulate nature under the assumed conditions. The following check list can be helpful in evaluating a model. First, the flow structure of the computer code should be carefully checked for possible programming errors. A useful test is to provide random numbers as initial input to each subroutine to check for indexing errors in the dimensional arrays. Second, the budgets of kinetic energy and of water should be computed to assure that the model conserves these quantities. Third, the results of a simulation should if possible be compared with those from a second model developed by independent researchers. Alternative forms of the same model can also be examined. Often sensitivity studies help in evaluating a particular model formulation. Finally, the model results should be compared with observations of natural clouds. A major difficulty in such comparisons is the large spatial and temporal variability of convection, caused by inhomogeneities in the atmospheric stratification and the boundary layer moisture supply, and interactions among clouds. These perturbations in the driving parameters are usually not recorded, and even if they were, it would be very difficult to use such measurements to provide initial and boundary data for the model. Furthermore, cumulus clouds usually coexist in varying sizes, intensities, and durations at any given moment. Therefore it is often appropriate, if not essential, to compare model results with statistical properties of an observed cloud population rather than with measurements from a single convective element. It is also appropriate to use cloud observations by remote sensors wherever possible, because they provide more complete coverage in space and time than in-situ measurements on the ground or in an airplane.

Figure 15.11 is an example of results from a careful comparison of a three-dimensional cloud ensemble simulation with an observed cloud

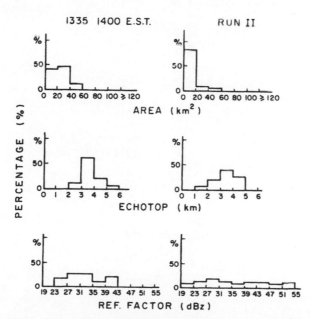

FIG. 15.11. Comparison of distributions of simulated and observed radar echo statistics of maximum area, maximum echo height, and maximum reflectivity factor obtained during the lifetime of a cell. (From Yau and Michaud, 1982.)

population. The radar echoes were analyzed at different times to obtain frequency distributions of the largest area of a cell, the maximum echo height and the peak reflectivity factor of a cell at any height. The figure compares the observed distributions with those obtained from the model. The distributions of area and peak reflectivity agree fairly well, while the correspondence in the distribution of echo height is good.

As a general comment about model evaluation, it should be stressed that the "science" of cloud modeling is still undergoing rapid development, and there are still basic uncertainties about the best choice of lateral boundary conditions, initialization procedures, and parameterization schemes for turbulence and ice phase processes. Therefore it is probably premature to derive either too much confidence or too much discouragement from a particular comparison between model results and nature. With the continuing development of larger and faster computers, and expanding research on cloud processes and model architecture, it is likely that future cloud simulations will become increasingly useful for the examination and understanding of convective storm processes.

References

ALMEIDA, F. C. DE (1976) The collisional problem of cloud droplets moving in a turbulent environment. Part I: A method of solution. *J. Atmos. Sci.* **33**, 1571–1578.

ALMEIDA, F. C. DE (1979) The effects of small-scale turbulent motions on the growth of a cloud droplet spectrum. *J. Atmos. Sci.* **36**, 1557–1563.

ATLAS, D. (ed.) (1989) *Radar Meteorology*. American Meteorological Society, Boston.

AUSTIN, P. M., and HOUZE, R. A. (1972) Analysis of the structure of precipitation patterns in New England. *J. Appl. Meteor.* **11**, 926–935.

AUSTIN, P. M., and KRAUS, M. J. (1968) Snowflake aggregation—a numerical model. *Proc. Intl. Conf. on Cloud Phys.*, Toronto, pp. 300–304.

BAKER, M. B., and LATHAM, J. (1979) The evolution of droplet spectra and the rate of production of embryonic raindrops in small cumulus clouds. *J. Atmos. Sci.* **36**, 1612–1615.

BATTAN, L. J. (1973) *Radar Observation of the Atmosphere*. University of Chicago Press, 324 pp.

BATTAN, L. J., and BRAHAM, R. R. (1956) A study of convective precipitation based on cloud and radar observations. *J. Meteor.* **13**, 587–591.

BATTAN, L. J., and THEISS, J. B. (1966) Observations of vertical motions and particle sizes in a thunderstorm. *J. Atmos. Sci.* **23**, 78–87.

BEARD, K. V. (1976) Terminal velocity and shape of cloud and precipitation drops aloft. *J. Atmos. Sci.* **33**, 851–864.

BEARD, K. V., and OCHS, H. T. (1984) Collection and coalescence efficiencies for accretion. *J. Geophys. Res.* **89**, 7165–7169.

BENNETTS, D. A., and HOSKINS, B. J. (1979) Conditional symmetric instability—a possible explanation for frontal rainbands. *Quart. J. Roy. Meteor. Soc.* **105**, 945–962.

BENNETTS, D. A., and SHARP, J. C. (1982) The relevance of conditional symmetric instability to the prediction of mesoscale frontal rainbands. *Quart. J. Roy. Meteor. Soc.* **108**, 595–602.

BERRY, E. X., and REINHARDT, R. L. (1974a) An analysis of cloud drop growth by collection: Part I. Double distributions. *J. Atmos. Sci.* **31**, 1814–1824.

BERRY, E. X., and REINHARDT, R. L. (1974b) An analysis of cloud drop growth by collection: Part II. Single initial distributions. *J. Atmos. Sci.* **31**, 1825–1831.

BOHREN, C. F. (1986) Cloud formation on descent revisited. *J. Atmos. Sci.* **43**, 3035–3037.

BOLTON, D. (1980) The computation of equivalent potential temperature. *Mon. Wea. Rev.* **108**, 1046–1053.

BOROVIKOV, A. M., KHRGIAN, A. KH., and others (1961) *Cloud Physics*. U.S. Dept. of Commerce, Washington, D.C., 392 pp. (Tr. from Russian by Israel Program for Scientific Translations.)

BOWEN, E. G. (1950) The formation of rain by coalescence. *Australian J. Sci. Res. A* **3**, 192–213.

BRAHAM, R. R. (1952) The water and energy budgets of the thunderstorm and their relation to thunderstorm development. *J. Meteor.* **9**, 227–242.

BRAHAM, R. R. (1968) Meteorological bases for precipitation development. *Bull. Amer. Meteor. Soc.* **49**, 343–353.

BRAZIER-SMITH, P. R., JENNINGS, S. G., and LATHAM, J. (1972) The interaction of falling water drops: coalescence. *Proc. Roy. Soc. Lond.* **A326**, 393–408.

BRAZIER-SMITH, P. R., JENNINGS, S. G., and LATHAM, J. (1973) Raindrop interactions and rainfall rates within clouds. *Quart. J. Roy. Meteor. Soc.* **99**, 260–272.

BROCK, J. R. (1972) Condensational growth of atmospheric aerosols. *Aerosols and Atmospheric Chemistry* (G. M. Hidy, Ed.), Academic Press, New York, 149–153.

BROWN, P. S. (1986) Analysis of the Low and List drop-breakup formulation. *J. Clim. Appl. Meteor.* **25**, 313–321.

BROWNING, K. A. (1964) Airflow and precipitation trajectories within severe local storms which travel to the right of the winds. *J. Atmos. Sci.* **21**, 634–639.

BROWNING, K. A., HARROLD, T. W., WHYMAN, A. J., and BEIMERS, J. G. D. (1968) Horizontal and vertical air motion and precipitation growth, within a shower. *Quart. J. Roy. Meteor. Soc.* **94**, 498–509.

BROWNING, K. A., and HARROLD, T. W. (1969) Air motion and precipitation growth in a wave depression. *Quart. J. Roy. Meteor. Soc.* **95**, 288–309.

BROWNING, K. A., and LUDLAM, F. H. (1962) Airflow in convective storms. *Quart. J. Roy. Meteor. Soc.* **88**, 117–135.

BUSHNELL, R. H. (1973) Dropsonde measurements of vertical winds in the Colorado thunderstorm of 22 July 1972. *J. Appl. Meteor.* **12**, 1371–1374.

BYERS, H. R. (1965) *Elements of Cloud Physics.* University of Chicago Press. 191 pp.

BYERS, H. R., and BRAHAM, R. R. (1949) *The Thunderstorm.* U.S. Dept. of Commerce, Washington, 287 pp.

CHARNEY, J. G. (1947) The dynamics of long waves in a baroclinic westerly current. *J. Meteor.* **4**, 135–163.

CHISHOLM, A. J. (1973) *Alberta Hailstorms. Part I: Radar Case Studies and Airflow Models.* Meteor. Monograph, Vol. 14, No. 36. American Meteorological Society, Boston, pp. 1–36.

CHISHOLM, A. J., and RENICK, J. H. (1972) Supercell and multicell Alberta hailstorms. *Abstracts, Intl. Conf. on Cloud Physics,* London, pp. 67–68.

COOPER, W. A., BAUMGARDNER, D., and DYE, J. E. (1986) Evolution of the droplet spectra in Hawaiian orographic clouds. *Preprints, Amer. Meteor. Soc. Conf. on Cloud Physics,* Snowmass, Colo., Vol. 2, 52–55.

DANIELSEN, E. F., BLECK, R., and MORRIS, D. A. (1972) Hail growth by stochastic coalescence in a cumulus model. *J. Atmos. Sci.* **29**, 135–155.

DENNIS, A. S. (1980) *Weather Modification by Cloud Seeding.* Academic Press, New York, 267 pp.

DOUGLAS, R. H., GUNN, K. L. S., and MARSHALL, J. S. (1957) Pattern in the vertical of snow generation. *J. Meteor.* **14**, 95–114.

DOUGLAS, R. H., and MARSHALL, J. S. (1954) The convection associated with the release of latent heat of sublimation. McGill University, Stormy Weather Group Rep. MW-19, 31 pp.

DOVIAK, R. J., and ZRNIC, D. S. (1984) *Doppler Radar and Weather Observations.* Academic Press, New York, 458 pp.

DRAKE, R. L. (1972a) A general mathematical survey of the coagulation equation. *Topics in Current Aerosol Research (Part 2)* (G. M. Hidy & J. R. Brock, Eds.), Pergamon Press, Oxford, 201–384.

DRAKE, R. L. (1972b) The scalar transport equation of coalescence theory: moments and kernels. *J. Atmos. Sci.* **29**, 537–547.

EADY, E. T. (1949) Long waves and cyclone waves. *Tellus* **1**, 33–52.

EAST, T. W. R. (1957) An inherent precipitation mechanism in cumulus clouds. *Quart. J. Roy. Meteor. Soc.* **83**, 61–76.

EMANUEL, K. A. (1979) Inertial instability and mesoscale convective systems. Part I: Linear theory of inertial instability in rotating viscous fluids. *J. Atmos. Sci.* **36**, 2425–2449.

EMANUEL, K. A. (1986) Overview and Definition of Mesoscale Meteorology. Ch. 1, pp. 1–17, in *Mesoscale Meteorology and Forecasting* (P. S. Ray, Ed.), American Meteorological Society, Boston.

FARLEY, R. D., and ORVILLE, H. D. (1986) Numerical modeling of hailstorms and hailstone growth. *J. Clim. Appl. Meteor.* **25**, 2014–2035.

FITZGERALD, J. W. (1972) A study of the initial phase of cloud droplet growth by condensation and comparison between theory and observation. Ph.D. thesis, University of Chicago, 144 pp.

FLETCHER, N. H. (1962) *The Physics of Rainclouds.* Cambridge University Press, 386 pp.

FRIEDLANDER, S. K. (1977) *Smoke, Dust and Haze.* Wiley, New York, 317 pp.

FRIEDMAN, R. M. (1982) Constituting the polar front, 1919–1920. *Isis* **73**, 343–362.
FUKUTA, N. (1969). Experimental studies on the growth of small ice crystals. *J. Atmos. Sci.* **26**, 522–531.
FUKUTA, N., and WALTER, L. A. (1970) Kinetics of hydrometeor growth from a vapor-spherical model. *J. Atmos. Sci.* **27**, 1160–1172.
GILLESPIE, D. T. (1972) The stochastic coalescence model for cloud droplet growth. *J. Atmos. Sci.* **29**, 1496–1510.
GILLESPIE, D. T. (1975) Three models for the coalescence growth of cloud drops. *J. Atmos. Sci.* **32**, 600–607.
GOSSARD, E. E., and STRAUCH, R. G. (1983) *Radar Observation of Clear Air and Clouds.* Elsevier, Amsterdam, 280 pp.
GUNN, K. L. S., and MARSHALL, J. S. (1958) The distribution with size of aggregate snowflakes. *J. Meteor.* **15**, 452–461.
GUNN, R., and KINZER, G. D. (1949) The terminal velocity of fall for water drops in stagnant air. *J. Meteor.* **6**, 243–248.
HALLETT, J., and MOSSOP, S. C. (1974) Production of secondary ice crystals during the riming process. *Nature* **249**, 26–28.
HARDY, K. R. (1963) The development of raindrop-size distribution and implications related to the physics of precipitation. *J. Atmos. Sci.* **20**, 299–312.
HARROLD, T. W., and AUSTIN, P. M. (1974) The structure of precipitation systems—a review. *J. Rech. Atmos.* **8**, 41–57.
HAVENS, B. S., JIUSTO, J. E., and VONNEGUT, B. (1978) *Early History of Cloud Seeding.* New Mexico Tech. Press, Socorro, 75 pp.
HEDLEY, M., and YAU, M. K. (1988) Radiation boundary conditions in numerical modeling. *Mon. Wea. Rev.* **116**, 1721–1736.
HEYMSFIELD, A. J., and MUSIL, D. J. (1982) The 22 July 1976 case study: Storm structure deduced from penetrating aircraft. In, *Hailstorms of the Central High Plains, Vol. II* (C. A. Knight and P. Squires, Eds.), Colorado Associated University Press, Boulder, pp. 163–180.
HOBBS, P. V. (1978) Organization and structure of clouds and precipitation on the mesoscale and microscale in cyclonic storms. *Rev. Geophys. Space Phys.* **16**, 741–755.
HOBBS, P. V. *et al.* (1974) The structure of clouds and precipitation over the Cascade Mountains and their modification by artificial seeding (1972–1973). Research Report 8, Department of Atmospheric Sciences, University of Washington, Seattle, 181 pp.
HOBBS, P. V., BOWDLE, D. A., and RADKE, L. F. (1985) Particles in the lower troposphere over the high plains of the United States. Part I: Size distributions, elemental composition and morphologies. *J. Clim. Appl. Meteor.* **24**, 1344–1356.
HOCKING, L. M., and JONAS, P. R. (1970) The collision efficiency of small drops. *Quart. J. Roy. Soc.* **96**, 722–729.
HOUGHTON, H. G. (1950) A preliminary quantitative analysis of precipitation mechanisms. *J. Meteor.* **7**, 363–369.
HOUGHTON, H. G. (1968) On precipitation mechanisms and their artificial modification. *J. Appl. Meteor.* **7**, 851–859.
HOUGHTON, H. G. (1985) *Physical Meteorology.* MIT Press, Cambridge, Mass., 442 pp.
HOUGHTON, H. G., and RADFORD, W. H. (1938) On the local dissipation of natural fog. *Papers in Phys. Ocean. and Meteor.*, Mass. Inst. of Technology and Woods Hole Ocean Inst. 6. No. 3, 63 pp.
HOUGHTON, J. T. (1986) *The Physics of Atmospheres* (2nd Ed.). Cambridge University Press, 271 pp.
HOUZE, R. A. (1981) Structures of atmospheric precipitation systems: A global survey. *Radio Sci.* **16**, 671–689.
JENSEN, B. J., and BLYTH, A. M. (1988) Comment on "Mixing mechanisms in cumulus congestus clouds. Part I: Observations". *J. Atmos. Sci.* **45**, 2460–2463.
JIUSTO, J. E., and WEICKMANN, H. K. (1973) Type of snowfall. *Bull. Amer. Meteor. Soc.* **54**, 1148–1162.
JONAS, P. R., and MASON, B. J. (1982) Entrainment and the droplet spectrum in cumulus clouds. *Quart. J. Roy. Meteor. Soc.* **108**, 857–869.

Joss, J., Thams, J. C., and Waldvogel, A. (1968) The variation of raindrop size distribution at Locarno. *Proc. Intl. Conf. on Cloud Phys.*, Toronto, pp. 369–373.

Kabanov, A. S., Mazin, I. P., and Smirvov, V. I. (1971) Comment on "Note on the theory of growth of cloud drops by condensation". *J. Atmos. Sci.* **28**, 129–130.

Kessler, E. (1969) *On the Distribution and Continuity of Water Substance in Atmospheric Circulations.* Met. Monograph, Vol. 10, No. 32, American Meteorological Society, Boston, 84 pp.

Klemp, J. B., and Wilhelmson, R. B. (1978a) The simulation of three-dimensional convective storm dynamics. *J. Atmos. Sci.* **35**, 1070–1096.

Klemp, J. B., and Wilhelmson, R. B. (1978b) Simulations of right and left moving storms through storm splitting. *J. Atmos. Sci.* **35**, 1097–1100.

Klemp, J. B., Wilhelmson, R. B., and Ray, P. S. (1981) Observed and numerically simulated structure of a mature supercell thunderstorm. *J. Atmos. Sci.* **38**, 1558–1580.

Klett, J. D., and Davis, M. H. (1973) Theoretical collision efficiencies of cloud droplets at small Reynolds numbers. *J. Atmos. Sci.* **30**, 107–117.

Knight, C. A. (1979) Ice nucleation in the atmosphere. *Adv. Colloid Interfac. Sci.* **10**, 369–395.

Knight, C. A., and Squires, P. (1982) *Hailstorms of the Central High Plains. Volume II: Case Studies of the National Hail Research Experiment.* Colorado Associated University Press, Boulder, 245 pp.

Koenig, L. R. (1971) Numerical modeling of ice deposition. *J. Atmos. Sci.* **28**, 226–237.

Komabayasi, M., Gonda, T., and Isono, K. (1964) Life times of water drops before breaking and size distribution of fragment droplets. *J. Meteor. Soc. Japan* **42**, 330–340.

Kropfli, R. A., and Miller, L. J. (1976) Kinematic structure and flux quantities in a convective storm from dual-Doppler radar observations. *J. Atmos. Sci.* **33**, 520–529.

Lamb, H. (1945) *Hydrodynamics.* Dover Publications, New York, 738 pp.

Langleben, M. P. (1954) The terminal velocity of snowflakes. *Quart. J. Roy. Meteor. Soc.* **80**, 174–181.

Lin, C. L., and Lee, S. C. (1975) Collision efficiency of water drops in the atmosphere. *J. Atmos. Sci.* **32**, 1412–1418.

List, R. J. (ed.) (1951) *Smithsonian Meteorological Tables.* Smithsonian Institution, Washington, 527 pp.

List, R., Donaldson, N. R., and Stewart, R. E. (1987) Temporal evolution of drop spectra to collisional equilibrium in steady and pulsating rain. *J. Atmos. Sci.* **44**, 362–372.

Long, A. B. (1971) Validity of the finite-difference droplet collection equation. *J. Atmos. Sci.* **28**, 210–218.

Low, R. D. H. (1969) A generalized equation for the solution effect in droplet growth. *J. Atmos. Sci.* **26**, 608–611.

Low, T. B., and List, R. (1982) Collision, coalescence and breakup of raindrops. Part I: Experimentally established coalescence efficiencies and fragment size distributions in breakup. *J. Atmos. Sci.* **39**, 1591–1606.

Macpherson, J. I., and Isaac, G. A. (1977) Turbulent characteristics of some Canadian cumulus clouds. *J. Appl. Meteor.* **16**, 81–90.

Manton, M. J. (1974) The collection kernel for coalescence of water droplets. *Tellus* **26**, 369–375.

Manton, M. J. (1979) On the broadening of a droplet distribution by turbulence near cloud base. *Quart. J. Roy. Meteor. Soc.* **105**, 899–914.

Marshall, J. S. (1953) Precipitation trajectories and patterns., *J. Meteor.* **10**, 25–29.

Marshall, J. S., and Palmer, W. McK. (1948) The distribution of raindrops with size. *J. Meteor.* **5**, 165–166.

Mason, B. J. (1971) *The Physics of Clouds.* Clarendon Press, Oxford, 671 pp.

Mason, B. J., and Jonas, P. R. (1974) The evolution of droplet spectra and large droplets by condensation in cumulus clouds. *Quart. J. Roy. Meteor. Soc.* **100**, 23–38.

Maxwell, J. C. (1890) *The Scientific Papers of James Clerk Maxwell, Vol. II* (W. D. Niven, Ed.). Cambridge University Press, pp. 636–640.

MAZIN, I. P. (1968) The stochastic condensation and its effect on the formation of cloud drop size distribution. *Proc. Intl. Conf. on Cloud Physics*, Toronto, pp. 67–71.

McDONALD, J. E. (1958) The physics of cloud modification. *Advances in Geophysics*, vol. 5, pp. 223–303, Academic Press Inc., New York.

McDONALD, J. E. (1963a) Early developments in the theory of the saturated adiabatic process. *Bull. Amer. Meteor. Soc.* **44**, 203–211.

McDONALD, J. E. (1963b) Use of the electrostatic analogy in studies of ice crystal growth. *Z. Angew. Math. Phys.* **14**, 610–620.

McTAGGART-COWAN, J. D., and LIST, R. (1975) Collision and breakup of water drops at terminal velocity. *J. Atmos. Sci.* **32**, 1401–1411.

MELZAK, A. Z., and HITSCHFELD, W. (1953) A mathematical treatment of random coalescence. McGill University, Stormy Weather Group Rep. MW-11, 28 pp.

MIDDLETON, W. E. K. (1966) *A History of the Theories of Rain*. Franklin Watts, Inc., New York, 223 pp.

MILLER, R. C., ANDERSON, R. J., KASSNER, J. L., and HAGEN, D. E. (1983) Homogeneous nucleation rate measurements for water over a wide range of temperature and nucleation rate. *J. Chem. Phys.* **78**, 3204–3211.

MORDY, W. (1959) Computations of the growth by condensation of a population of cloud droplets. *Tellus* **11**, 16–44.

MORGAN, J. J. (1982) Factors governing the pH, availability of H^+, and oxidation capacity of rain. *Atmospheric Chemistry* (E. D. Goldberg, Ed.), pp. 17–40. Dahlem Konferenzen 1982. Springer-Verlag, Berlin.

MOSSOP, S. C. (1985) The origin and concentration of ice crystals in clouds. *Bull. Amer. Meteor. Soc.* **66**, 264–273.

NAKAYA, U. (1954) *Snow Crystals*. Harvard University Press, 521 pp.

NICHOLLS, S. (1984) The dynamics of stratocumulus: aircraft observations and comparisons with a mixed-layer model. *Quart. J. Roy. Meteor. Soc.* **110**, 783–820.

NICHOLLS, S., and TURTON, J. D. (1986) An observational study of the structure of stratiform cloud sheets. Part II. Entrainment. *Quart. J. Roy. Meteor. Soc.* **112**, 461–480.

NOONKESTER, V. R. (1984) Droplet spectra observed in marine stratus cloud layers. *J. Atmos. Sci.* **41**, 829–845.

OCHS, H. T. (1978) Moment-conserving techniques for warm cloud microphysical computations. Part II: Model testing and results. *J. Atmos. Sci.* **35**, 1959–1973.

OGURA, Y., and PHILLIPS, N. A. (1962) Scale analysis of deep and shallow convection in the atmosphere. *J. Atmos. Sci.* **19**, 173–179.

PALUCH, I. R. (1979) The entrainment mechanism in Colorado cumuli. *J. Atmos. Sci.* **36**, 2467–2478.

PALUCH, I. R., and KNIGHT, C. A. (1984) Mixing and the evolution of cloud droplet size spectra in a vigorous continental cumulus. *J. Atmos. Sci.* **41**, 1801–1815.

PALUCH, I. R., and KNIGHT, C. A. (1986) Does mixing promote cloud droplet growth? *J. Atmos. Sci.* **43**, 1994–1998.

PANCHEV, S. (1971) *Random Functions and Turbulence*. Pergamon Press, Oxford, 444 pp.

PANOFSKY, H. A. (1981) Atmospheric hydrodynamics. In *Dynamic Meteorology, an Introductory Selection* (B. W. Atkinson, Ed.). Methuen, London, 228 pp.

PITTER, R. L., and PRUPPACHER, H. R. (1974) A numerical investigation of collision efficiencies of simple ice plates colliding with supercooled water drops. *J. Atmos. Sci.* **31**, 551–559.

PRUPPACHER, H. R., and KLETT, J. D. (1978) *Microphysics of Clouds and Precipitation*. Reidel, Dordrecht, Holland, 714 pp.

REUTER, G. W., VILLIERS, R. DE, and YAVIN, Y. (1988) The collection kernel for two falling cloud drops subjected to random perturbations in a turbulent air flow: a stochastic model. *J. Atmos. Sci.* **45**, 765–773.

REUTER, G. W., and YAU, M. K. (1987a) Mixing mechanisms in cumulus congestus clouds. Part I. Observations. *J. Atmos. Sci.* **44**, 781–797.

REUTER, G. W., and YAU, M. K. (1987b) Mixing mechanisms in cumulus congestus clouds. Part II. Numerical simulations. *J. Atmos. Sci.* **44**, 798–827.

REUTER, G. W., and YAU, M. K. (1987c) Numerical modelling of cloud development in a sheared environment. *Beitr. Phys. Atmos.* **60**, 65–80.

RIGBY, E. C., MARSHALL, J. S., and HITSCHFELD, W. (1954) The development of the size distribution of raindrops during their fall. *J. Meteor.* **11**, 362–372.

ROBERTSON, D. (1974) Monte Carlo simulations of drop growth by accretion. *J. Atmos. Sci.* **31**, 1344–1350.

ROGERS, R. R. (1967) Doppler radar investigation of Hawaiian rain. *Tellus* **19**, 432–455.

ROGERS, R. R. (1984) A review of multiparameter radar observations of precipitation. *Radio Sci.* **19**, 23–36.

ROGERS, R. R., and SAKELLARIOU, N. K. (1986) Precipitation production in three Alberta thunderstorms. *Atmosphere–Ocean* **24**, 145–168.

ROGERS, R. R., and SMITH, P. L. (1983) Radar meteorology. *Sci. Prog. Oxf.* **68**, 149–176.

ROTUNNO, R. (1986) Tornadoes and Tornadogenesis. Ch. 18, pp. 414–436, in *Mesoscale Meteorology and Forecasting* (P. S. Ray, Ed.). American Meteorological Society, Boston.

RYAN, B. T. (1974) Growth of drops by coalescence: the effect of different collection kernels and of additional growth by condensation. *J. Atmos. Sci.* **31**, 1942–1948.

RYAN, B. T., BLAU, H. H., THUNA, P. C., COHEN, M. L., and ROBERTS, G. D. (1972) Cloud microstructure as determined by optical cloud particle spectrometer. *J. Appl. Meteor.* **11**, 149–156.

SCHEMENAUER, R. S., MACPHERSON, J. I., ISAAC, G. A., AND STRAPP, J. W. (1980) Canadian participation in HIPLEX 1979. Report APRB 110 P 34, Atmospheric Environment Service, Environment Canada, 206 pp.

SCHLAMP, R. J., GROVER, S. N., PRUPPACHER, H. R., and HAMIELIC, A. E. (1976) A numerical investigation of the effect of electric charges and vertical external electric fields on the collision efficiency of cloud drops. *J. Atmos. Sci.* **33**, 1747–1755.

SCHLESINGER, R. E. (1982) Effects of mesoscale lifting, precipitation and boundary layer shear on severe storm dynamics in a three-dimensional numerical modeling study. *Preprints, 12th Conference on Severe Local Storms*, pp. 536–541. American Meteorological Society, Boston.

SCHNELL, R. C., and VALI, G. (1976) Biogenic ice nuclei: Part I. Terrestrial and marine sources. *J. Atmos. Sci.* **33**, 1554–1564.

SCHWARTZ, S. E. (1987) Both sides now: the chemistry of clouds. *Environmental Sciences* (F. S. Sterrett, Ed.), pp. 89–144. Annals New York Acad. Sci. **502**.

SCORER, R. S. (1958) *Natural Aerodynamics*. Pergamon Press, Oxford, 312 pp.

SCORER, R. S., and WEXLER, H. (1963) *A Colour Guide to Clouds*. Pergamon Press, Oxford, 63 pp.

SCOTT, W. T. (1968) On the connection between the Telford and kinetic equation approaches to droplet coalescence theory. *J. Atmos. Sci.* **25**, 871–873.

SCOTT, W. T. (1972) Comments on "Validity of the finite-difference droplet collection equation". *J. Atmos. Sci.* **29**, 593–594.

SEKHON, R. S., and SRIVASTAVA, R. C. (1970) Snow size spectra and radar reflectivity. *J. Atmos. Sci.* **27**, 299–307.

SELTZER, M. A, PASSARELLI, R. E., and EMANUEL, K. A. (1985) The possible role of symmetric instability in the formation of precipitation bands. *J. Atmos. Sci.* **42**, 2207–2219.

SLINN, W. G. N. (1975) Atmospheric aerosol particles in surface-level air. *Atmospheric Environment* **9**, 763–764.

SMIRNOV, V. I., and NADEYKINA, L. A. (1984) Analytical solution to a kinetic equation for the cloud drop size spectrum formed by condensation in a turbulent medium. *Proc. 9th Intl. Conf. on Cloud Physics*, Tallinn, Vol. I, pp. 233–235.

SMOLARKIEWICZ, P. K., and CLARK, T. L. (1985) Numerical simulation of the evolution of a three dimensional field of cumulus clouds. Part I: Model description, comparison with observations and sensitivity studies. *J. Atmos. Sci.* **42**, 502–522.

SQUIRES, P. (1958a) The microstructure and colloidal stability of warm clouds. *Tellus* **10**, 256–271.

SQUIRES, P. (1958b) Penetrative downdrafts in cumuli. *Tellus* **10**, 381–389.

SRIVASTAVA, R. C. (1967) A study of the effect of precipitation on cumulus dynamics. *J. Atmos. Sci.* **24**, 36–45.

SRIVASTAVA, R. C. (1969) Note on the theory of growth of cloud drops by condensation. *J. Atmos. Sci.* **26**, 776–780.

STEINER, J. T. (1973) A three-dimensional model of cumulus cloud development. *J. Atmos. Sci.* **30**, 414–435.

STEINER, M., and WALDVOGEL, A. (1987) Peaks in raindrop size distributions. *J. Atmos. Sci.* **44**, 3127–3133.

SUMMERS, P. W. (1982) From emission to deposition: processes and budgets. Report AQRB-82-005-D, Atmospheric Environment Service, Environment Canada, 35 pp.

TAKEDA, T. (1971) Numerical simulation of a precipitating convective cloud: the formation of a "long-lasting" cloud. *J. Atmos. Sci.* **28**, 350–376.

TELFORD, J. W. (1955) A new aspect of coalescence theory. *J. Meteor.* **12**, 436–444.

TELFORD, J. W. (1987) Comments on "Does mixing promote cloud droplet growth?" *J. Atmos. Sci.* **44**, 2352–2354.

TELFORD, J. W., and CHAI, S. K. (1980) A new aspect of condensation theory. *Pure Appl. Geophys.* **118**, 720–742.

TELFORD, J. W., KECK, T. S., and CHAI, S. K. (1984) Entrainment at cloud top and the droplet spectra. *J. Atmos. Sci.* **41**, 3170–3179.

TELFORD, J. W., and WAGNER, P. B. (1981) Observations of condensation growth determined by entity type mixing. *Pure Appl. Geophys.* **119**, 934–965.

THOMSON, W. (1870) On the equilibrium of vapour at a curved surface of liquid. *Proc. Roy. Soc. Edinb.* **7**, 63–68.

TURPEINEN, O., and YAU, M. K. (1981) Comparison of results from a three-dimensional cloud model with statistics of radar echoes on day 261 of GATE. *Mon. Wea. Rev.* **109**, 1495–1511.

TWOMEY, S. (1959) The nuclei of natural cloud formation: the supersaturation in natural clouds and the variation of cloud droplet concentration. *Geofis. Pura et Appl.* **43**, 243–249.

TWOMEY, S. (1964) Statistical effects in the evolution of a distribution of cloud droplets by coalescence. *J. Atmos. Sci.* **21**, 553–557.

TWOMEY, S. (1966) Computations of rain formation by coalescence. *J. Atmos. Sci.* **23**, 405–411.

VALDEZ, M. P., and YOUNG, K. C. (1985) Number fluxes in equilibrium raindrop populations: a Markov chain analysis. *J. Atmos. Sci.* **42**, 1024–1036.

VALI, G. (1985) The origin and concentration of ice crystals in clouds. *Bull. Amer. Meteor. Soc.* **66**, 264–273.

VALI, G., CHRISTENSEN, M., FRESH, R. W., GALYAN, E. L., MAKI, L. R., and SCHNELL, R. C. (1976) Biogenic ice nuclei. Part II. Bacterial sources. *J. Atmos. Sci.* **33**, 1565–1570.

WALDVOGEL, A. (1974) The N_0 jump in raindrop spectra. *J. Atmos. Sci.* **31**, 1067–1078.

WALLACE, J. M., and HOBBS, P. V. (1977) *Atmospheric Science, An Introductory Survey.* Academic Press, New York, 467 pp.

WARNER, J. (1969a) The microstructure of cumulus cloud. Part I. General features of the droplet spectrum. *J. Atmos. Sci.* **26**, 1049–1059.

WARNER, J. (1969b) The microstructure of cumulus cloud. Part II. The effect of droplet size distribution of the cloud nucleus spectrum and updraft velocity. *J. Atmos. Sci.* **26**, 1272–1282.

WARSHAW, M. (1967) Cloud droplet growth: statistical foundations and a one-dimensional sedimentation model. *J. Atmos. Sci.* **24**, 278–286.

WEXLER, A. (1976) Vapor pressure formulation for water in range 0 to 100°C. A revision. *J. Res. Nat. Bur. Stand.* **80A**, 775–785.

WEXLER, A. (1977) Vapor pressure formulation for ice. *J. Res. Nat. Bur. Stand.* **81A**, 5–20.

WEXLER, R. (1955) An evaluation of the physical effects in the melting layer. *Proc. Fifth Wea. Radar Conf.*, Ft. Monmouth, N.J., pp. 329–334.

WEXLER, R. (1960) Efficiency of natural rain. *Physics of Precipitation* (H. Weickmann, Ed.), Geophys. Monograph 5, American Geophysical Union, pp. 158–163.

WILHELMSON, R. B., and CHEN, C. S. (1982) A simulation of the development of successive cells along a cold outflow boundary. *J. Atmos. Sci.* **39**, 1466–1483.

WILHELMSON, R. B., and KLEMP, J. B. (1978) A numerical study of storm splitting that leads to long-lived storms. *J. Atmos. Sci.* **35**, 1974–1986.

WILHELMSON, R. B., and KLEMP, J. B. (1981) A three-dimensional numerical simulation of splitting severe storms on 3 April 1964. *J. Atmos. Sci.* **38**, 1581–1600.

WOODCOCK, A. H., DUCE, R. A., and MOYERS, J. L. (1971) Salt particles and raindrops in Hawaii. *J. Atmos. Sci.* **28**, 1252–1272.

YAU, M. K. (1979) Perturbation pressure and cumulus convection. *J. Atmos. Sci.* **36**, 690–694.

YAU, M. K., and MICHAUD, R. (1982) Numerical simulation of a cumulus ensemble in three dimensions. *J. Atmos. Sci.* **39**, 1062–1079.

YOUNG, K. C. (1975) The evolution of drop spectra due to condensation, coalescence and breakup. *J. Atmos. Sci.* **32**, 965–973.

Appendix

The 1964 ICAO Standard Atmosphere
(Z = altitude above mean sea level in meters)

Z	gpm	Temperature, K	Pressure, kPa	Density, kg/m^3
0	0	288.150	101.325	1.2250
200	200	286.850	98.945	1.2017
400	400	285.550	96.611	1.1786
600	600	284.250	94.322	1.1560
800	800	282.951	92.077	1.1337
1000	1000	281.651	89.876	1.1117
1200	1200	280.351	87.718	1.0900
1400	1400	279.052	85.602	1.0687
1600	1600	277.753	83.528	1.0476
1800	1799	276.453	81.494	1.0269
2000	1999	275.154	79.501	1.0066
2500	2499	271.906	74.692	0.9570
3000	2999	268.659	70.121	0.9093
3500	3498	265.413	65.780	0.8634
4000	3997	262.166	61.660	0.8194
4500	4497	258.921	57.753	0.7770
5000	4996	255.676	54.048	0.7364
5500	5495	252.431	50.539	0.6975
6000	5994	249.187	47.218	0.6601
6500	6493	245.943	44.075	0.6243
7000	6992	242.700	41.105	0.5900
7500	7491	239.457	38.300	0.5572
8000	7990	236.215	35.652	0.5258
8500	8489	232.974	33.154	0.4958
9000	8987	229.733	30.801	0.4671
9500	9486	226.492	28.585	0.4397
10,000	9984	223.252	26.500	0.4135
10,500	10,483	220.013	24.540	0.3886
11,000	10,981	216.774	22.700	0.3648
11,500	11,479	216.650	20.985	0.3374
12,000	11,977	216.650	19.399	0.3119
12,500	12,475	216.650	17.934	0.2884
13,000	12,973	216.650	16.580	0.2666
13,500	13,471	216.650	15.328	0.2465
14,000	13,969	216.650	14.170	0.2279
14,500	14,467	216.650	13.101	0.2107
15,000	14,965	216.650	12.112	0.1948
15,500	15,462	216.650	11.198	0.1801
16,000	15,960	216.650	10.353	0.1665
16,500	16,457	216.650	9.572	0.1539
17,000	16,955	216.650	8.850	0.1423
17,500	17,452	216.650	8.182	0.1316
18,000	17,949	216.650	7.565	0.1217
18,500	18,446	216.650	6.995	0.1125
19,000	18,943	216.650	6.468	0.1040
19,500	19,440	216.650	5.980	0.0962
20,000	19,937	216.650	5.529	0.0889

Answers to Selected Problems

Chapter 1

1.1. Each constituent gas separately obeys the ideal gas law, $p_n = R^* M_n T/m_n V$, where p_n, M_n, and m_n denote, respectively, the pressure, mass, and molecular weight of the nth constituent. The volume occupied by the mixture is V. By Dalton's law, the total pressure is the sum of the partial pressures, $p = \Sigma p_n = (R^* T/V) \Sigma (M_n/m_n)$. Therefore $pa = R^* T \Sigma (M_n/m_n)/\Sigma M_n$ and $\bar{m}$, the effective mean molecular weight of the mixture, is given by

$$\bar{m} = \Sigma M_n/\Sigma (M_n/m_n).$$

1.3. (a) $n = 1.23$; (b) $\Delta u = -1.30 \times 10^4$ J; (c) $\Delta w = 2.25 \times 10^4$ J; (d) $\Delta q = 9.5 \times 10^3$ J.

1.6. 308 K.

Chapter 2

2.1. A thermodynamic diagram for water with coordinates of e versus α shows that isothermal compression of the vapor leads to condensation if $T < T_{CP}$, the temperature of the critical point (647 K). A thermodynamic diagram with coordinates of e versus T can be used to show that adiabatic expansion of the vapor leads to condensation if $T < L/c_{pv}$. Both of these inequalities are satisfied in the atmosphere.

2.2. (a) $\theta = 285.8$ K; (b) $w = 2.11$ g/kg; (c) $T_d = -10.6°C$; (d) $T_c = -12°C$; (e) $T_w = -7°C$; (f) $\theta_w = 277.5$ K; (g) $T_e = 0°C$; (h) $T_v = -4.7°C$; (i) $\varrho = 1.04$ kg/m³.

2.3. Note that the wet-bulb temperature remains constant in this process. For any increase Δw in mixing ratio, we can therefore calculate the decrease in temperature of the air from (2.29) and the new saturation mixing ratio from (2.18), where $e = e_s(T)$ and we use (2.12) for e_s. By iteration, we find that $\Delta w = 3.7 \times 10^{-3}$ leads to a relative humidity of 65%. Taking account of the mass of air in the room, we conclude that 0.48 kg of water must be evaporated.

Chapter 3

3.1. From the definition, the top of the homogeneous atmosphere is at the height $h = p_0/\varrho_0 g$, where p_0 is the surface pressure. The surface pressure p of the exponential atmosphere is given by g times the integral of density over height, and equals $\varrho_0 g H$. Therefore $h = H$ if $p = p_0$.

3.3. Let h denote the geometric altitude above the earth with radius R. Then, by the inverse square law,

$$\text{gpm}(z) = \frac{\psi(z)}{g_0} = \frac{1}{g_0} \int_0^z g(h)dh = \frac{1}{g_0} \int_0^z g_0 \left(\frac{R}{R+h}\right)^2 dh$$

$$= \frac{z}{1 + z/R} \approx z\left(1 - \frac{z}{R}\right) = z - az^2,$$

where
$$a = 1/R = 1.57 \times 10^{-7} \text{ m}^{-1}.$$

Therefore at geometric altitudes of 1, 10, and 50 km, the heights in geopotential units are 0.9998, 9.9843, and 49.608 km.

3.5. 373 K.

Chapter 4

4.2. At the dew point, $e_s(T_d) = e$. Therefore the dew point varies with e according to
$$\frac{de}{e} = \frac{\varepsilon L}{R' T_d^2} dT_d.$$

When the surface parcel is lifted with w held constant,
$$\frac{de}{e} = \frac{dp}{p} = \frac{1}{p}(-\varrho g)dz = \frac{\varepsilon L dT_d}{R' T_d^2}.$$

Therefore
$$\frac{dT_d}{dz} = -\frac{g T_d}{\varepsilon L},$$

and in adiabatic ascent,
$$\frac{d}{dz}(T - T_d) = -\Gamma + \frac{g T_d}{\varepsilon L} = -\Gamma\left(1 - \frac{c_p T_d}{\varepsilon L}\right).$$

For $T_d \approx 270$ K,
$$\frac{d}{dz}(T - T_d) \approx -0.83\Gamma \approx -8 \text{ K/km}.$$

Therefore, H (in km) $\approx (T - T_d)/8$.

4.3. In general,
$$\frac{1}{f}\frac{\partial f}{\partial z} = \frac{1}{w}\frac{\partial w}{\partial z} + \frac{L\varepsilon\gamma}{R' T^2} - \frac{\Gamma c_p}{R' T}.$$

When well mixed,
$$\frac{1}{f}\frac{\partial f}{\partial z} = \frac{\Gamma}{R' T}\left(\frac{L\varepsilon}{T} - c_p\right) > 0$$

because $T < L\varepsilon/c_p$.
    For $T = 250$ K and $f = 50\%$, $\partial f/\partial z = 3.6\%$ per 100 m.

4.6. (a) $\theta = 31.5°C$; $\varrho = 1.09$ kg/m^3; $w = 11.7$ g/kg; $f = 57\%$; $T_v = 25.5°C$; $T_e = 54°C$;
    $\theta_e = 66°C$; $T_w = 17.5°C$; $\theta_w = 21°C$; $T_c = 12.5°C$.
    (b) $p_c = 795$ mb; $\chi = 7.1$ g/kg; $\Delta\phi = 61.8$ J kg^{-1} K^{-1}; $Q = 17.8$ kJ/kg; $U = 35$ m/s.
    (c) $W = 2.25$ cm.

4.7. $\alpha = -1/4$; $\beta = -5/4$.

Chapter 5

5.1. (a) $M = 0.84$ g/m^3; (b) $\varrho_v = 9.4$ g/m^3; (c) $\varrho = 0.985$ kg/m^3; (d) mean distance = 0.21 cm.

5.3. 0.85 g/m^3 for the trade-wind cumulus; 0.47 g/m^3 for the continental cumulus.

5.4. Denoting the mean radius by $\bar{r}$, the concentration by N, and the cloud water content by M, we have

$$B = 3/\bar{r}, \quad A = NB^3/2 = \frac{27N}{2\bar{r}^3}; \quad \text{and} \quad A = \frac{1.45 \times 10^{-6}M}{2\bar{r}^3}.$$

5.5.

$T(^\circ C)$	-10	-10.5	-10
$p\,(\text{kPa})$	60	60	60
$M\,(\text{g/m}^3)$	1	1	1.2
$Q\,(\text{g/kg})$	4.19	4.08	4.45
$\theta_q\,(K)$	310.8	309.9	310.8

Chapter 6

6.1. $k = 4.5 \times 10^{-8} \text{ cm}^3 \text{ min}^{-1}$.

6.2. The mass M of sodium chloride per unit mass of water decreases as $M^{-1/2}$. At 280 K, this ratio decreases from 1.04×10^{-1} at 10^{-19} g to 3.3×10^{-4} at 10^{-14} g.

6.4. Number density $= 1.38 \times 10^5 \text{ cm}^{-3}$; surface area $= 1.29 \times 10^{-4} \text{ cm}^2/\text{cm}^3$.

6.7. The number of nuclei activated at supersaturation s is the integral of $n_d(D)$ over all nuclei for which $s^* < s$. But from equations 6.7 and 6.8, $s^* \propto D^{-3/2}$, where D is the diameter of the nucleus whose critical supersaturation is s^*. Integration over the Junge distribution gives the answer indicated.

6.10. $r^* = 0.075\,\mu\text{m}; T = 293 \text{ K}$.

Chapter 7

7.2. $s = 0.45\%$.

7.4. $q_1 \approx -L/R_v T^2; q_2 \approx p/\varepsilon e_s$.

7.7. From (7.29) and problem 7.5, the limiting value of supersaturation is given by

$$s_\infty = \frac{\omega}{\eta} = G(p, T)U/(\nu_0 \bar{r})$$

where

$$G(p, T) = \frac{100Q_1(F_k + F_d)}{4\pi\varrho_L Q_2}.$$

(Note that r in the expression for η has been replaced by $\bar{r}$, the mean drop radius, when considering droplets having a range of sizes.) For the conditions given in this problem, $G(p, T) = 1.78 \times 10^2 \%\text{s/kg}$ and $\nu_0 = 6.84 \times 10^8 \text{ kg}^{-1}$, as determined from the third moment of the Gaussian distribution. Therefore $s_\infty = 0.52\%$.

Chapter 8

8.1. (a) 0.35 mm radius; (b) 20.9 min; (c) 0.38 mm radius.

8.2. Including ventilation, the drops with initial radii of 1, 0.5, and 0.1 mm require times of 27.7, 11.5, and 1.06 min to evaporate. Neglecting ventilation, the times are increased to 167, 41.7, and 1.67 min.

8.4. (a) 0.99 mm diameter.

(b) The final radius is determined by solution of the equation

$$\int_{r_0}^{r_f} \left(\frac{U-u}{u}\right) dr = \int_{z_0}^{z_f} \frac{EM}{4\varrho_L}\, dz = 0.$$

A graphical solution, obtained by plotting $1/u$ versus r and integrating, gives $r_f = 1.6\,\text{mm}$.

(c) 16.7 min.

8.6. Hint: the complete solution, for condensation and accretion acting together, is

$$r^2(t) = \left(r_0^2 + \frac{\xi}{b}\right)e^{2bt} - \frac{\xi}{b},$$

where $b = EMk_3/4\varrho_L$ and $\xi = (S-1)/(F_k + F_d)$.

8.7. Time constant $= u/g$.

Chapter 9

9.1. Time $= 60.4\,\text{min}$; distance of fall $= 1.26\,\text{km}$.

9.2. $\lambda = 0.987$; $T_0 = -44.7°\text{C}$.

Chapter 10

10.2. Time required is 5.58 min, during which the particle ascends 3.10 km.

10.3. Change in cloud water content due to sweepout is given by

$$\left(\frac{dM}{dt}\right)_{acc} = -\frac{3}{2}\,\pi E N_0 k M R^{4/3}(4\pi N_0 k)^{-4/5}.$$

Condensation rate is given by

$$\left(\frac{dM}{dt}\right)_{cond} = \varrho^2 U \Gamma_s \varepsilon L w_s/pT.$$

Balance exists when

$$\left(\frac{dM}{dt}\right)_{acc} + \left(\frac{dM}{dt}\right)_{cond} = 0.$$

For typical conditions the balance implies, numerically and in CGS units,

$$(U/M)R^{-4/5} = 9 \times 10^{10}E.$$

Note that the condensation rate, following from (7.22), may also be obtained from

$$\left(\frac{dM}{dt}\right)_{cond} = \varrho\,\frac{d\chi}{dt} = \varrho\,\frac{Q_1}{Q_2}\,U,$$

which is equivalent to the expression given earlier.

Chapter 11

11.1. $Z = 6!(4\pi k)^{-7/5}N_0^{-2/5}R^{7/5}$.

For Z in mm^6/m^3 and R in mm/h, $Z = 216R^{1.4}$.

11.2. (a) $Z = 180R^{1.56}$.

(b) $R = \dfrac{\pi}{6} N_0 3! \left[\dfrac{A}{b^4} - \dfrac{B}{(b + C)^4} \right]$,   where $b = \left(\dfrac{N_0 6!}{Z} \right)^{1/7}$.

11.4. At time t the reflectivity factor is given by

$$Z(t) = 2^6 \int_0^\infty n(r, t) r^6 dr.$$

Droplet growth is described by $r(t) = (r_0^2 + \alpha)^{1/2}$, where $\alpha = 2t(S - 1)/(F_k + F_d)$. In terms of the initial droplet spectrum, $n(r, t) = n_0(r_0)(dr_0/dr)$. Therefore

$$Z(t) = 2^6 \int_0^\infty n_0(r_0)(r_0^2 + \alpha)^3 dr_0$$
$$= 2^6 N[\overline{r^6} + 3\alpha \overline{r^4} + 3\alpha^2 \overline{r^2} + \alpha^3],$$

where N is the droplet concentration and $\overline{r^j}$ is the mean jth power of radius in the initial distribution. Since $Z(0) = 2^6 N \overline{r^6}$, it follows that

$$\frac{Z(t)}{Z(0)} = 1 + (3\alpha \overline{r^4} + 3\alpha^2 \overline{r^2} + \alpha^3)/\overline{r^6}. \tag{A}$$

The even moments of a Gaussian distribution are

$$\overline{r^2} = \sigma^2 + (\overline{r})^2; \qquad \overline{r^4} = 3\sigma^4 + 6\sigma^2(\overline{r})^2 + (\overline{r})^4;$$
$$\overline{r^6} = 15\sigma^6 + 45\sigma^4(\overline{r})^2 + 15\sigma^2(\overline{r})^4 + (\overline{r})^6.$$

Substituting these into (A) and employing $\sigma = 0.15r$ gives

$$\frac{Z(t)}{Z(0)} = 1 + (3.41\alpha(\overline{r})^4 + 3.07\alpha^2(\overline{r})^2 + \alpha^3)/(\overline{r})^6. \tag{B}$$

For the conditions given, $F_k + F_d \approx 0.14 \times 10^7$ s/cm^2. Consequently, the reflectivity at time t, measured in decibels above the initial value, is

$$10 \log \frac{Z(t)}{Z(0)} = 19.3 \text{ dB at 5 min}$$
$$= 26.6 \text{ dB at 10 min}$$
$$= 31.3 \text{ dB at 15 min.}$$

Notice that for a monodisperse distribution ($\sigma \to 0$),

$$\frac{Z(t)}{Z(0)} = 1 + (3\alpha r^4 + 3\alpha^2 r^2 + \alpha^3)/r^6,$$

which is a very good approximation to the more accurate result (B).

11.5. $Z = Z_0(N_0/N)$ and $R = R_0(N_0/N)^{1/3}$ so that $(Z/Z_0) = (R/R_0)^3$.

Chapter 12

12.1. So defined, the sweepout efficiency of a particle equals the product of its velocity times its cross-section. Thus, $E(\text{graupel})/E(\text{rain}) = 0.48m^{-2/15}$ where m is the particle mass. It can be further shown that graupel has a higher sweepout efficiency than a raindrop of equal mass if the graupel radius $r_g < 0.2$ cm. Conversely, if $r_g > 0.2$ cm the raindrop has the higher efficiency.

12.2. The reflectivity factor at time t is given by

$$Z(t) = 2^6 \int_0^\infty n(r, t)r^6 dr.$$

But the drop-size distribution evolves according to (12.3). Therefore

$$Z(t) = 2^6 \int_0^\infty r^6 e^{-at} n_0(re^{-at})dr$$

which reduces to

$$Z(t) = e^{6at} Z(0).$$

The signal increases by 10 dB when

$$t = \ln 10/6\alpha \approx 77 \text{ s}.$$

12.5. Rain rate is given by

$$R = \frac{\pi}{6} \int_0^\infty N(D)D^3 V(D)dD = \pi N_0 k 4!/6b^5.$$

Rate of change of cloud water due to rain-collection is

$$\frac{dM}{dt} = -\frac{\pi}{4} \int_0^\infty D^2 N(D)V(D)EMdD.$$

Therefore

$$\frac{dM}{M} = -\frac{3}{2} E(\pi k N_0)^{1/5} 4^{-4/5} R^{4/5} dt.$$

For $E \approx 1$ and R expressed in consistent units, this leads to the required result.

12.6. Hint: when mass of snow dm per unit mass of air melts, the air temperature changes by dT and the saturation specific humidity changes by dq_s. From the first law of thermodynamics, these changes are related by

$$L_f dm + c_p dT + L dq_s = 0,$$

where L_f is the latent heat of fusion. The amount of water condensed per unit mass of snow that melts is $-dq_s/dm$.

12.7

$$\bar{R} = \left(\frac{\alpha}{\alpha - 1} \right) \frac{R_{max}^{1-\alpha} - R_{min}^{1-\alpha}}{R_{max}^{-\alpha} - R_{min}^{-\alpha}}.$$

Chapter 13

13.1. (a) 10 mm/h; (b) 1.39 TW; (c) 2.5 m/s; (d) Evaporation would not affect the net rate of latent heating but would require a stronger vertical velocity to produce the same amount of rain at the ground.

13.3. $f_{ke} = 1.09 \times 10^{-1}$ W/m².

13.5. The heat balance implies that

$$REML_f u(R) = 4K(T_s - T_0 - \gamma h)b$$

where γ is the lapse rate and h is the distance of fall from 5 km, where the ambient temperature is T_0. The ventilation factor may be shown to be of the form $b = k_2 R^{3/4}$, where $k_2 \approx 50$. The growth rate is given by

$$dR/dh = EM/4\varrho_s.$$

Solving for EM from the balance relation, substituting into the growth equation and integrating, leads to the indicated result.

13.6. The equilibrium sizes for different temperatures are as follows:

$T(°C)$	0	−2	−4	−6	−8	−10	−12	−14
$R(\text{mm})$	5.6	4.2	3.0	2.1	1.4	0.77	0.32	0.04

13.7. The rate of change of temperature of the hailstone is given by

$$dT_s/dt = -(mc_s)^{-1}dQ_s/dt,$$

where m and c_s denote its mass and specific heat. From (13.3) this equation may be written

$$dT_s/dt + \alpha T_s = \beta$$

where $\alpha = 4\pi RKb/mc_s$ and $\beta = \alpha T$.
   The solution of this equation is

$$(T_s - T) = (T_0 - T)e^{-\alpha t},$$

where $T_0 = T_s(0)$. Therefore the time constant τ is given by $\tau = 1/\alpha = R^2\varrho_s c_s/3Kb$. For the conditions specified the time constant amounts to approximately 26 s.

Chapter 14

14.1. The depletion of cloud water by accretion is given by

$$\left(\frac{dM}{dt}\right)_{\text{acc}} = -\pi EMD^2NV(D)/4.$$

The condensation rate may be determined in various ways (see solution to problem 10.3) and is given approximately by

$$\frac{1}{U}\left(\frac{dM}{dt}\right)_{\text{cond}} = 5 \times 10^{-12} \quad (\text{CGS units}).$$

Under the conditions given, the balance requires that the concentration N of small hailstones must exceed approximately 100 per m^3, or 0.1 per liter.

14.2. After glaciation, $T = -18.7°C$.

Index

A-scope, radar, 185
Absolute humidity, defined, 16
Accommodation coefficient, 113–114
Accretion, 163, 206
Acidic precipitation, 218–220
Activity spectrum of condensation nuclei,
95–96, 144–145
Adiabatic liquid water content, 23, 25
Adiabatic process
defined, 6
significance in atmosphere, 6
Aerodynamic resistance, effect on
convection, 54
Aerosols, 82, 86, 89–90, 154, 219
coagulation, 93, 96
defined, 1
mass concentrations, 91
number concentrations, 90
removal by precipitation, 93
sedimentation, 93
size distribution functions, 91–94
accumulation mode, 93
coarse particle mode, 93
Junge distribution, 92, 96
nucleation mode, 93
typical shapes, 94
those important in natural cloud
formation, 94–95
Aggregation, 163, 206
Aitken nucleus counter, 97
Aitken particles, 90
Ammonia, 219
Ammonium sulfate, 87–88, 95
Anelastic assumption, 248
Antenna gain, 187
Argon, 1
Autoconversion, 141, 250
Avogadro's Law, 2

Backscatter (radar) cross-section
defined, 188
of small sphere, 189
Baroclinic instability, 41, 196
Beamwidth, radar, 186
Bergen school, 209
Bergeron, Tor, 209
Bezold, Wilhelm von, 21
Bjerknes, Vilhelm, 209

Boiling point, 42
Bowen, E. G., 131, 156
Bowen meteor hypothesis, 156
Bowen model, 131–133
Brunt–Väisälä frequency, 31–32, 196, 214
Bubble theory, 54–55
Buoyancy force, 30, 39, 50

Capacitance of ice crystals, 159
CAPPI display, 192
Carbon dioxide, 1, 218
Clausius–Clapeyron equation, 14–15, 20,
45, 81
approximate integrated form, 14
Cloud chemistry, 218–220
Cloud condensation nuclei, 73
defined, 95
Cloud droplets
measured sizes and concentrations, 69
sweepout by precipitation, 217, 221
Cloud modeling
basic equations, 248
computing requirements, 253
model evaluation, 266–267
simulation of cloud ensembles, 263
simulation of mesovortex, 263
simulation of storm splitting, 260–262
Cloud-top mixing, 70–71, 123, 259
Clouds
aircraft measurements in, 65–70, 73, 75
classification, 62–63
humidity fluctuations within, 73–74
likelihood of containing ice, 75, 156
likelihood of containing precipitation, 75
measured droplet sizes and
concentrations, 69
microphysical properties of, 66–68
microstructure, 64
patterns observed by satellite, 60–62
scales of variability, 62–64
significance in chemical processes, 70
stability of microstructure, 82
stratus, microstructure of, 74–75
tendency for supercooling, 151
turbulence in, 67–68
variation of properties with height, 65,
68–69
Coagulation coefficient, 137

Coalescence, 82–83, 121–122, 163
  stimulation by cloud seeding, 241
Coalescence efficiency, 124
  of raindrops, 173–174, 177
Coefficient of ionic activity, 87
Coefficient of thermal conductivity, 100
  table, 103
Collection efficiency, 124, 130
Collision efficiency, 124, 126
  defined, 127
  linear, defined, 129
  plotted, 128, 130
  table, 129
Columnar water vapor (precipitable water),
  42
Condensation coefficient, 113–114, 117
Condensation nuclei, 81–82, 86
  activation of, defined, 89
  activity spectrum, 95
  atmospheric, 89–90
  chemical composition, 90, 94
  classification by size, 90
  sizes and shapes, 90
Condensed water, effect on parcel
  buoyancy, 50
Conditional instability, 32, 196
Continuity equation, 229
Contributing volume, radar, 188
Convection, elementary theory, 48–49
Convective condensation level, 47–48
Convective instability, 33–35
Coriolis force, 35–37, 39
Critical radius for homogeneous nucleation,
  85
Critical radius of solution droplet, 89
Cyclic process, 4–5

Dew point temperature, 19
Diffusion coefficient of water vapor in air, 99
  table, 103
Diffusion equation, 99
Doppler
  effect, 193
  radar, 193
  spectrum, 193
Downdraft
  formation in thunderstorm, 225
  penetrative, 53, 59, 70–71
Drag coefficient, 124
Drag force, on falling sphere, 124
Drizzle drops, defined, 105
Drop evaporation, effect on distance of fall,
  105
Drop-size distribution of rain
  analytical approximation, 171
  dependence on rain rate, 170
  equilibrium form, 179

examples, 171
  tendency for exponential form, 172, 175,
    179
Droplet growth
  by collision-coalescence, 124, 147–148
    effects of turbulence, 145–146
    statistical effects, 134
  by condensation and coalescence, 143–145
  by continuous collection, 130–131
  by discrete captures, 135–137
  by molecular diffusion, 99–101
    kinetic corrections, 112–115
    Mason's analytical approximation, 102
    nonstationary growth, 116–117
    statistical effects, 118
    ventilation effects, 116
  stochastic coalescence, 137–139
  trajectories in updraft, 132–133
Droplet growth by condensation
  approximation for steady supersaturation,
    104
  correlation between supersaturation and
    vertical velocity, 117
  effect of unsteady updraft, 117
  growth of droplet populations, 105–109
    evolution of the droplet spectrum, 110
    limiting supersaturation and relaxation
    time, 110
  stochastic effects, 118–119
  ventilation effects, 148
Droplet spectrum
  approximations for, 72, 79
  comparison of computed and measured,
    115
  differences in continental and maritime
    clouds, 72–73
  distribution function defined, 72
  evolution by condensation, 110–111
    approximate equation, 111
  examples of aircraft measurements, 69
  form near cloud base, 112
  general characteristics, 72
  question of spectral broadening, 112,
    117–119, 122–123
  tendency for bimodality, 72, 117
Dry ice, 240, 242, 245

Enthalpy, 11
Entrainment, 52, 70–71, 122–123, 225
Entrainment rate, 53
Entropy
  defined, 7
  of cloudy air, 25
  relation to potential temperature, 8
Equation of state
  dry air, 1
  moist air, 18

perturbation expansion, 247
  water vapor, 12
Equivalent potential temperature
  defined, 23
  vertical profile in rainband, 216
Equivalent temperature, 20
  adiabatic definition, 23
Exact differential, 4–5, 9

Fall speed
  aerosol, 97
  cloud droplets, 125
  graupel, 164
  hail, 238
  ice crystals, 164
  of water drops:
    approximate formulas, 125–126
    table, 126
  raindrops, 125
  snowflakes, 164
First law of thermodynamics, 2
Fog, formation by mixing of air masses, 45,
  57
Free energy barrier, 81
Free energy of crystal/liquid interface,
  150–151
Free energy of crystal/vapor interface, 151

Gamma probability distribution, 140, 143
Gas constant
  dry air, 2
  individual, 2
  moist air, 18
  universal, 2
  water vapor, 12
General dynamic equation, 111
General Electric Research Laboratory, 240
Geopotential
  defined, 41
  relation to geometric altitude, 42
Geostrophic wind, 36–37, 214
Gibbs function, 13
Graupel, 83
  fall speed, 164

Hail, 83
  equilibrium temperature, 237
  fall speed, 238
  growth theory, 235–237
  suppression by cloud seeding, 243–244
Hail embryos, 244
Hawaiian orographic rain, radar pattern, 202
Heat diffusion from a growing droplet, 100
Hertz, Heinrich, 21
Heterogeneous nucleation

of ice, 151
  experimental procedures, 153
  threshold temperatures for different
    substances, 154
  of liquid, 81
Hodograph, 260
Homogeneous atmosphere, defined, 42
Homogeneous deposition, 151
Homogeneous freezing, temperature
  threshold, 151
Homogeneous nucleation
  of ice, 150
  of liquid, 82, 84–86
Howard, Luke, 62–63
Hutton, James, 45
Hydrogen peroxide, 219
Hydrostatic equation, 28
  perturbation expansion, 247
Hygrometric chart, 44–45
Hygroscopic nuclei, 86

Ice
  likelihood of occurrence in clouds, 75, 156
  nucleation, 150
  promotion by cloud seeding, 242
Ice crystal growth, significance of diffusion,
  150
Ice crystal habit, dependence on
  temperature and humidity, 163
Ice crystal multiplication, 157–158
Ice crystal process, 83
Ice crystals
  concentration in clouds, 156
    dependence on temperature, 157
  discrepancy between crystal and nucleus
    concentrations, 156–157
  fall speeds, 164
  fracturing caused by collisions, 158
  growth by accretion, 163
    approximate equation, 165
  growth by aggregation, 165
    approximate equation, 166
  growth by molecular diffusion, 158
    approximate equation, 160
    dependence of growth rate on pressure
      and temperature, 161
    electrostatic analogy, 159–160
    kinetic effects, 162
    rate compared to coalescence, 166–168
    ventilation effects, 160
  growth habit, 162
  relation between mass and size, 165
Ice nuclei, 82
  biogenic, 155–156
  chemical composition, 155
  methods of sampling, 154

Ice nuclei—*continued*
   threshold temperatures, 155
   typical atmospheric concentrations, 155
Instabilities, atmospheric, length and time
   scales, 196
Internal energy, 3, 5
Isentropic condensation point, 22
Isentropic condensation temperature, 19–21
Isentropic process, 8
Isentropic surface, 40–41
Isobaric process, 6
Isochoric process, 6
Isothermal process, 6

Joule's Law, 3
Junge, Christian, 92

Kaolinite, 155
Kelvin, Lord, 21, 85
Kelvin's equation, 84, 87
Kessler parameterizations, 250, 253
Knudsen number, 112
Köhler curve, 88

Langmuir, Irving, 127, 240
Lapse rate
   ambient, 30
   dry adiabatic, 29
   modification by convection, 47–48
   pseudoadiabatic, 32
   superadiabatic, 47
Latent heat of fusion, 15
Latent heat of sublimation, 14–15
Latent heat of vaporization, 13
   dependence on temperature, 14–15
   tabulated, 16
Lifting condensation level, 48

Marshall–Palmer distribution, 171–172, 191
Maxwell, James Clerk, 100, 159
Maxwell theory of droplet growth, 100, 113
Mean free path, 112
Melting layer (bright band)
   condensation within, 221
   explanation, 200
   radar examples, 198, 199
Mesocyclone, 62, 234
   simulation in cloud models, 263
Mixing
   adiabatic, 46–47
   dilution of parcel by, 52
   effect on droplet spectrum, 117, 122–123
   effect on parcel buoyancy, 52–53
   entity-type, 123
   evidence of, from measurements in
      clouds, 68–71

homogeneous, 122
inhomogeneous, 123
isobaric, 44–46
Mixing ratio
   defined, 16
   total water, 46
   defined, 25
Moments of an exponential distribution, 180

Nitrogen, 1
Normand's rule, 23
   illustrated, 22
Nucleation
   heterogeneous, 81
   homogeneous, 85–86
   ice, 150–151
      different mechanisms, 152–153
      experimental uncertainties, 153
   rate of formation of droplet embryos, 86
Nucleation of water in vapor, homogeneous,
   84

Oxidation, 219
Oxygen, 1
Ozone, 1, 219

Paluch diagram, 70–71, 79
Parcel theory
   elementary, 29–31, 48
   modifications to elementary theory, 50
   vertical velocity predicted by, 49
Penetrative downdrafts, 70–71
Perturbation pressure, 247, 250
Plume theory, 55–56
Poisson probability distribution, 134, 139
Poisson's equation for adiabatic process, 7
Polarization of radar waves, 193
Polytropic process, 11
Positive area of sounding, 50
Potential temperature
   defined, 7
   graphical determination, 22
PPI display, 191–192
   examples
      chaotic rain pattern, 212
      line of showers, 210
      mesoscale rain pattern, 211
      mesoscale rainband, 215
      showers, 204
      snow and melting layer, 198–199
      tornado, 232
Precipitation
   initiation by ice-crystal process or
      coalescence, 168, 170
   likelihood of occurrence in clouds, 75–77

mesoscale structure, 210–213, 215–216
 formation of rainbands by symmetric
  instability, 214
Precipitation efficiency, 217
Precipitation production, by thunderstorms,
 234–235
Precipitation rate, in terms of particle size
 distribution, 182
Precipitation theories, 206–208
 continuous-rain approximation, 208
 shower approximation, 208
Precipitation water content
 defined, 183
 rate of development in shower, 209
Pressure gradient force, 36–37, 39
Pseudoadiabatic process, 21–22
Pseudomonas syringae (bacterium), 156

Radar equation, 187–190
Radar parameters, typical values, 186
Rain, time required for formation, 121
Raindrop breakup
 character of satellite drops, 177
 collision-induced, 173–176
  effect on drop-size distribution, 179
  kinetic energy considerations, 176–177
  production of satellite drops, 173
  size distributions of satellite drops, 178
 spontaneous, 172
Raoult's Law, 87
Rayleigh scattering, 189
Receiver noise, radar, 191
Reflectivity factor
 defined, 190
 on scale of dBz, 190–191
 relation to precipitation rate, 190–191
Refraction, of radar waves, 186
Relative humidity, defined, 17
Reversible saturated adiabatic process, 25
Reynolds number, 116, 125
RHI display, 192
 examples, snow and melting layer, 198
Richardson number, 214, 216
Rime-splintering, 158

Satellite drops, from raindrop breakup, 178
Saturation ratio
 critical value for solution droplet, 88–89
 defined, 85
Saturation vapor pressure
 accurate empirical formula, 16
 of ice, 15
 of water, 14
 tabulated, 16

Sea salt, 94
 importance as giant nuclei, 122
Second law of thermodynamics, 7
Showers, 203
 radar pattern, 204–207
Signal fluctuations, radar, 188
Silver iodide, 153, 240, 242–244
Slantwise displacement, 38
Slice method of stability analysis, 51
Snow generating cells, 200
Snowflakes
 aggregation and breakup, 182
 distribution with size, 180
  exponential approximation, 181
 fall speed, 164, 202
 formation by aggregation, 166
 formation of, 83
Sodium chloride, 87, 89
Solution droplet, vapor pressure of, 87–88
Specific heat capacity
 defined, 5
 of dry air, 5
 of hailstone, 239
 of liquid water, 14
  dependence on temperature, 15
 of moist air, 18
 of water vapor, 14
  dependence on temperature, 15
Specific humidity, 46
 defined, 17
Spongy ice, 237
Stability criteria
 dry air, 30–31
 moist air, 32
Stochastic coalescence equation, 138–139
 examples of solutions, 140–142
Stochastic condensation, 118
Stokes' Law, 97, 125
Substrate, for ice nucleation, 151
Sulfate, 219
Sulfuric acid, 219
Supercooling, 82, 151–152
Supersaturation, 82
 maximum expected for a given nucleus
  activity spectrum, 96
 rate of change in cloudy updraft, 106
 relative to ice in a water saturated
  environment, 161
Surface tension
 defined, 85
 dependence on temperature, 98
 value for water, 85
Symmetric instability, 39–40, 196, 216

Telford model, 134–136
Thermal gradient diffusion chamber, 97
Thermal wind, defined, 38

Thermodynamic charts
  emagram, 8
    illustrated, 9
  Stüve diagram, 8
    illustrated, 8
  tephigram, 9
    illustrated, 10, 24
Thunderstorm, radar structure, 227
Thunderstorms
  aircraft measurements in, 77–79
  cloud microstructure of, 77, 79
  life cycle, 222–225
  multicell, 228–229
    illustrated, 231
  precipitation production by, 234–235
  radar structure, 226, 228
  supercell, 226–228
    illustrated, 230
  updraft structure, 232
  vertical velocity profiles, 225
Time constant
  for droplet response to turbulence, 149
  for hailstone response to temperature
    changes, 238
Tornado
  conditions attendant to formation, 234
  photograph, 233
Trace gases, 1, 219
True thermodynamic diagram, defined, 8

Van't Hoff factor, 87
Vapor pressure
  defined, 16
  of solution droplet, 87–88

reduction by dissolved material, 86
saturation, of water and ice, tabulated, 16
Ventilation effects, 116
  on ice crystal growth, 160
Vertical air velocity in clouds, 66–67
Vertical velocity, from Doppler
  measurements, 229
Virtual temperature, defined, 17
Viscosity
  dynamic, 97
    approximate formula, 102
    table, 103
  kinematic, 127
Vorticity, 40, 214

Wake capture, 128
Washington, University of, 214
Water content of clouds, 66–68, 70, 74–75,
  79
Water vapor, 1
  equation of state, 12
Wet equivalent potential temperature, 26,
  46, 59, 70
Wet-bulb potential temperature
  defined, 23
  graphical determination, 22
  use in stability analysis, 34
Wet-bulb temperature, 20
  adiabatic definition, 23
Work done in expansion, 3–5

Zeldovich factor, 86